Reading the Weather

by Thomas Morris

I0484817

DEDICATED with love, to my grandmother MARY GIBSON HALDEMAN herself responsible for so much sunshine.

CONTENTS

FORECAST

Science is certainly coming into her own nowadays,--and into everybody else's. Every activity of man and most of Nature's have felt her quickening hand. Her eye is upon the rest. Drinking is going out because the drinker is inefficient. The fly is going out because he carries germs. And for everything that goes out something else comes in that makes people healthier and more comfortable, and, perhaps, wiser.

One strange thing about this flood-tide of science is that it overwhelms the old, buttressed superstitions the easiest of all, once it really sets about it. For instance, nothing could have been better fortified for centuries than the fact that night air is injurious and should be shut out of house. Then, science turned its eye upon night air, found it a little cooler, a trifle moister, and somewhat cleaner than day air with the result that we all invite it indoors, now, and even go out to meet it.

Once interested in the air, science soon began to take up that commonplace but baffling phase of it called the weather. Now, of all matters under the sun the weather was the deepest intrenched in superstition and hearsay. From the era of Noah it had been made the subject of more remarks unrelieved by common sense than any other. It was at once the commonest topic for conversation and the rarest for thought. Considering the opportunities for study of the weather this conclusion, we must admit, is more surprising than complimentary to the human race. But it is so. The fact that science had to face was this: that the weather had been and remained a tremendous, dimly-recognized factor in our level of living. So talk about it all must. And science set about finding some easy fundamental truths to talk instead of the hereditary gossip about old-fashioned winters or the usual meaningless conversational coin.

Two groups of men had always known a good deal about the weather from experience: the sailor had to know it to save his life, and the farmer had to cultivate a weather eye along with his early peas. But the ordinary business man (and wife), the town-dweller, and even the suburbanite knew so few of the proven facts that the weather from day to day, from hour to hour, was a continual puzzle to them. The rain not only fell upon the just and unjust but it fell unquestioned, or misunderstood.

At last Science established some sort of a Weather Bureau in 1870, in our country, and after this had triumphed over great handicaps, the Government set it upon its present footing in 1891. An intelligent interest in the weather was in likelihood of being aroused by maps, pamphlets, frost and flood warnings that saved dollars and lives. Then suddenly, or almost suddenly, a new force was felt in every community. It was the call of outdoors. The new land of woods and lakes was explored. Men learned that living by bread alone (without air) made a very stuffy existence. Hence the man in town opened all his windows at night, the suburban majority planned to build sleeping porches, the youngsters begged to go to camp, their fathers went hunting and fishing in increasing numbers, and, most important of all, the fathers' wives began to accompany them into the woods.

Thus, living has been turned inside out,--the very state of things that old scientist Plato recommended some thirty thousand moons ago. And among the manifestations of nature the weather is holding its place, important and even fascinating. For the person who most depends on umbrellas and the subway in the city needs to watch the sky most carefully in the woods. That old academic question as to whether it be wise or foolish to come in out of the wet was never settled by the wilderness veteran. The veteran's wife settles it very quickly. She considers the cloud. When the commuter goes camping he rightly likes his comforts. A wet skin is not one of these. Therefore he studies the feel of the wind.

And so it comes about that the person who talks about storm centers and areas of high pressure and cumulus clouds is no longer regarded as slightly unhinged. Men are eager to learn the laws of the snowstorm and the cold wave; for, with the knowledge that snow is not poison and cold not necessarily discomfort, January has been opened up for enjoyments that July could never give.

Bookwriting and camping are both explained by the same fact,--a certain fondness for the thing. I wanted to see the commoner weather pinned down to facts. The following chapters resulted. They constitute a sort of Overhead Baedeker, it being their pleasure to show up the sureties of the sun and rain and to star the weather signs that can be relied upon. For, after all, even the elements, although unruled, are law-abiding.

READING THE WEATHER

CHAPTER I

OUR WELL-ORDERED ATMOSPHERE

If there is anything that has been overlooked more than another it is our atmosphere. But it absolutely cannot be avoided--in books on the weather. It deserves a chapter, anyway, because if it were not for the atmosphere this earth of ours would be a wizened and sterile lump. It would float uselessly about in the general cosmos like the moon.

To be sure the earth does not loom very large in the eye of the sun. It receives a positively trifling fraction of the total output of sunheat. So negligible is this amount that it would not be worth our mentioning if we did not owe our existence to it. It is thanks to the atmosphere, however, that the earth attains this (borrowed) importance. It is thanks to this thin layer of gases that we are protected from that fraction of sunheat which, however trifling when compared with the whole, would otherwise be sufficient to fry us all in a second. Without this gas wrapping we would all freeze (if still unfried) immediately after sunset. The atmosphere keeps us in a sort of thermos globe, unmindful of the burning power of the great star, and of the uncalculated cold of outer space.

Yet, limitless as it seems to us and inexhaustible, our invaluable atmosphere is a small thing after all. Half of its total bulk is compressed into the first three and a half miles upward. Only one sixty-fourth of it lies above the twenty-one mile limit. Compared with the thickness of the earth this makes a very thin envelope.

Light as air, we say, forgetting that this stuff that looks so thin and inconsequential weighs fifteen pounds to the square inch. We walk around carrying our fourteen tons gaily enough. The only reason that we don't grumble is because the gases press evenly in all directions permeating our tissues and thereby supporting this crushing burden. A layer of water thirty-four feet thick weighs just about as much as this air-pack under which we feel so buoyant. But if these gases get in motion we feel their pressure. We say the wind is strong to-day.

As it blows along the surface of the earth this wind is mostly nitrogen, oxygen, moisture, and dust. The nitrogen occupies nearly eight-tenths of a given bulk of air, the oxygen two-tenths, and the moisture anything up to one-twentieth. Five other gases are present in small quantities. The dust and the water vapor occupy space independently of the rest. As one goes up mountains the water vapor increases for a couple of thousand feet and then decreases to the seven mile limit after which it has almost completely vanished. The lightest gases have been detected as high up as two hundred miles and scientists think that hydrogen, the lightest of all, may escape altogether from the restraint of gravity. One strange fact about all of these gases is that they do not form a separate chemical combination, although they are thoroughly mixed.

At first glance the extreme readiness of the atmosphere to carry dust and bacteria does not seem a point in its favor. In reality it is. Most bacteria are really allies of the human race. They benefit us by producing fermentations and disintegrations of soils that prepare them for plant food. It is a pity that the few disease breeding types of bacteria should have given the family a bad name. Without bacteria the sheltering atmosphere would have nothing but desert rock to protect.

Further, rain is accounted for only by the dust. Of course this sounds very near the world's record in absurdities. But it is a half truth at least, for moisture cannot condense on nothing. Every drop of rain, every globule of mist must have a nucleus. Consequently each wind that blows, each volcano that erupts is laying up dust for a rainy day. Apparently the atmosphere is empty. Actually it is full enough of dust-nuclei to outfit a fullgrown fog if the dewpoint should be favorable. If there were no dust in the air all shadows would be intensest black, the sunlight blinding.

But the dust particles fulfill their greatest mission as heat collectors,--they and the particles of water vapor which have embraced them. It is in reality owing to these water globules and not to the atmosphere that supports them that we are enabled to live in such comfortable temperatures. For the air strata above seven miles where the tides of oxygen and nitrogen have rid themselves of water and dust absorb very little of the solar radiation. The heat is grabbed by the lowest layer of air as it goes by. The air snatches it both going and coming. The little particles get about half of it on the way down and when it is radiated back very little escapes them.

So it comes about that the heavy moist air near the earth is the warmest of all. It would, of course, get very warm if, as it collected its heat, it didn't have a tendency to rise. As it rises, moreover, it must fight gravity, that arch enemy of all rising things. And as it fights it loses energy, which is heat. So high altitudes and low temperatures are found together for these two reasons. But after the limit of moisture content has been reached the temperature gets no lower according to reliable investigations. Instead a monotony of 459?below zero eternally prevails--459?is called the absolute zero of space.

The vertical heating arrangements of the atmosphere appear somewhat irregular. But horizontally it is in a much worse way. The surface of the globe is three quarters water and one quarter land and irregularly arranged at that. The shiny water surfaces reflect a good deal of the heat which they receive, they use up the heat in evaporation and what they do absorb penetrates far. The land surfaces, on the contrary, absorb most of the heat received, but it does not penetrate to any depth. As a consequence of these differences land warms up about four times as quickly as water and cools off about four times as fast. Therefore the temperature of air over continents is liable to much more rapid and extreme changes than the air over the oceans.

The disparity of temperature is also rendered much greater because of differing areas of cloud and clear skies, because of interfering mountain masses, because of the change from day to night, or the constant progress of the seasons. At first blush it seems remarkable that the atmosphere should not be hopelessly unsettled in its habits, that there should belong to it any hint of system. As a matter of fact, in the main its courses are as well-ordered as the sun's. Cause and not caprice are at the bottom of the wind's listings. Its one desire is rest.

Cirrus clouds first appear as feathery lines converging toward one or two points on the horizon, often merging into bands of darker clouds, arranged horizontally. A sky like this appears when there is little wind. If the wind shifts to an easterly direction by way of north there will likely be snow within 24 hours; if it works around by way of the southwest and south 36 hours will probably pass before rain. If the mares' tails, as here, are absent and yet the stratified clouds are present there is little likelihood of a storm. Cirrus clouds precede every disturbance of magnitude. Sometimes they are hidden by a lower cloud layer.]

But rest it rarely succeeds in finding. Forever warming, rising, cooling, falling, it rushes about to regain its equilibrium. With so many opposing forces at work the calm day is the real marvel, our weeks of Indian Summer the ranking miracle of our climate. The very evolution of the myriad patches of air quilted over the earth with their different opportunities to become heated, to cool their heels, precludes stability in our so called Temperate Zone. But over great stretches of the earth's surface conditions are continuous enough to discipline the atmosphere into strict routine. Conjure the globe before your eyes and you will find the scheme of atmospheric circulation something like this:

A broad band of heated air perpetually rises from the sweltering equatorial belt of lands and seas. The supply never ceases, the warming process goes on night and day, and to a great height the light warm incense mounts. Then, cooling, from this altitude it begins to run down hill toward the poles. This is happening all the way around the globe. So naturally the common centers, the poles, cannot accommodate all this downrush of air. Therefore as it approaches the goal it falls into a majestic file about the center, very much as water does in running out of a hole in the center of a circular basin. The nearer north, the cooler this vast maelstrom grows and the nearer has it sunk to the earth. It descends circuitously and, by the force arising from the earth's rotation, is sheered to the right in the northern hemisphere, to the left in the southern.

Watching the water circle out of the basin you will notice the outside whirl is in no hurry to get to the center. This corresponds to the easterly trades of commerce, geography, and fiction. The direction of the upper currents flowing back to the poles is from southwest to northeast; but in our middle zones this becomes almost from west to east, is constant and is known to the profession as the prevailing westerlies.

Look up some day when wisps of clouds are floating very high. You will notice that their port is in the east, mattering not what wind may be blowing where you are. They are above the petty disturbances of the shallow surface winds. They follow a Gulf Stream of immeasurable grandeur. Onward, always onward, they sail, emblems of a great serenity.

Beneath this vast drift of air, which increases in velocity as it nears the pole, is an undertow from two to three miles thick. It is the movements of this undertow that affect our lives. These movements are influenced by all the changes of temperature and by the configurations of land. They take the form of whirls. These whirls may be small eddies, local in effect, or vast cyclones with diameters of fifteen hundred miles. Small or large they roll along under the Westerlies, translated by friction, and invariably moving for most of their course in an easterly direction, like their tractor above. They circle across the United States every few days. Their courses do not vary a great deal, and yet enough to make each one a matter for conjecture. And all the conjecturing centers upon the condition of the atmosphere,--the changing atmosphere which is yet so dependable.

The weather we are used to, the daily weather that catches us unprepared, and yet that does not mistreat us all the time is the product of these little whirls, which are so remotely connected with the grander atmospheric movements of our planet. Remembering this, we can at last come back to earth and set about our real business which is to see why certain kinds of weather come at such uncertain times and how to tell when they will arrive.

CHAPTER II

THE CLEAR DAY

We owe our fair weather to that department of atmospheric activity called anticyclone by the weatherman. The anticyclone is an accumulation of air which has become colder than the air surrounding it. This accumulation oftener than not has an area near the center where the air is coldest. About this coldest area the air currents revolve in the direction of a clock's hands. And since this cold air is contracted and denser than its warmer environment it has a perpetual tendency to whirl outward from the center into this warmer environment.

One comes to think, therefore, of the anticyclone as a huge pyramid of cold air moving slowly across the country from west to east and all the while melting down on all sides, like a plate of ice-cream, into the surrounding territory. It is such an immense accumulation that often while its head is reared over Montana the first shivers of its approach are beginning to be felt in Texas and Pennsylvania. It does not extend equally far, however, to the north and west of its head, which is really sometimes where its tail ought to be. That is, a long slope of increasing pressure and cold will sweep in a gentle gradient from Pennsylvania to Montana and will then decrease by a very steep gradient to the Pacific Coast.

The anticyclone draws its power from the inexhaustible supplies of cold air from the upper levels. This air is very dry and accounts for the almost invariably clear skies of the anticyclone.

In winter when the intensity of all the atmospheric activities is greatly increased, the anticyclone develops into the cold wave. The rapidly rising pressure rears its head and rushes along upon the heels of a storm like a vast tidal wave at sixty miles an hour, tumbling the mercury thirty, forty, fifty degrees.

These cold waves first appear in the northwest. They cannot well originate over either ocean and a high-pressure area building up over the southern half of the country will not attain the sufficient degree of frigidity to earn the title, for even cold waves have been standardized by the Government. But although nearly all the cold waves choose Montana or the Dakotas as a base, they have at least two definite lines of action. Those which are born amid the mountains or on the great plains of Montana have a curious habit of bombarding the Texas coast before starting on their eastward march. It is not unusual for us to read of zero weather in the Panhandle and freezing on the Gulf while the mercury may still be standing as high as fifty in New York City.

It is this rapid onslaught from Montana to Texas that produces those notorious blizzards of that section called northers, during which the cattle used to be frozen on the hoof. The record time for a drive of this extent is about twelve hours and the normal about twenty-four which gives scant time for the Weather Bureau to warn the vast interests of the impending assault. When the cold wave, after following this path, does swing toward the Atlantic Coast, as most of them do, it has lost interest and usually produces only seasonably cold weather along the Appalachians.

Those cold waves that recruit their strength in Canada and enter the United States through Minnesota or, rarely, this side of the Lakes move along the border and supply intensely cold weather for a night or two to New England and the Middle Atlantic States.

Cold waves almost always follow a storm. The storm, being an area of low pressure makes a fit receptacle for the surplus of the high pressure, and since the whole business of the weather is to seek peace and pursue it, the greater the discrepancies the more violent the pursuit. Consequently we have the spectacle of a ridge of cold dry air following and trying to level up a fleeing hollow of warm moist air--but rarely succeeding. This principle of action and reaction is almost the sole principle of the weather and is nowhere more clearly demonstrated than in the winter's succession of storm and cold wave.

In summer the anticyclones are not only actually but relatively more moderate than in winter. But their influence is still the same,--clear skies, cooler nights, dry, westerly winds. During the year the anticyclone furnishes us with about sixty per cent. of our weather. The cyclone is responsible for the remaining forty per cent. The weather depends on the cyclone for its variety and upon the anticyclone for its reputation. So it is well to be able to recognize an anticyclone when one appears.

The first and most reliable symptom of the approach of an anticyclone is the west wind. This sign is valid the country over, and is one of the very few signs that hold true for most of the North Temperate Zone. In summer over our country the west wind comes from the southwest, to be Irish, and in winter from the northwest. But for nearly all of our forty-eight states for nearly all of the year the westerly winds are those that bring us fair days and nights. And it is these crisp, clear days and cloudless, brilliant nights which we have in mind when we boast to English friends of our American weather.

The west wind is so popular because it has a slight downward flowing tendency. It also blows from land to sea over all America except the narrow Pacific coast. These downward, outward directions allow it to gather only enough moisture to keep it from becoming seriously dry. Its upper sources supply it with ozone. Its density gives it weight and by its superior weight it prevails. It dries roads faster than a brace of suns could do it. It is tonic. And curiously enough, although the anticyclone loads half a ton excess weight upon us we like it. The greater the burden the more we feel like leaping and shouting. Our good cheer seems to be ground out of us, like street pianos.

The reverse holds, too. For when the anticyclone moves off us and the cyclone hovers over us, removing half a ton of pressure, instead of feeling relieved we feel depressed, out of spirit. The animals share this reaction with us. In fact barnyards antedated barometers as forecasters, because all the domestic creatures, with pigs in particular, evidenced the disagreeable leniency of the low pressure areas upon their persons.

"Grumphie smells the weather An' Grumphie smells the wun' He kens when clouds will gather An' smoor the blinkin' sun."

The only trouble about this rather extravagant tribute to the pig, versatile though he is, is that he can tell only a very few hours ahead about the coming changes and it takes so much more skill to judge what his actions mean than to read the face of the sky that the science of meteorology finally comes to supplant barnyardology.

The coming of the anticyclone is foretold by the shifting of the wind from any quarter to the west. The course that the center of the anticyclone is keeping may be watched by the same agency. Since the circulation from the cone of cold air follows the hour hands of a clock it follows that if the center is moving north of you the wind, blowing outward from the center, will work from west to northwest and from northwest to north and slightly east of north.

If the wind has shifted into the west on a Wednesday, it will likely be cold by Wednesday night and colder on Thursday. By Friday morning the wind will be coming from the north, likely, with the lowest temperature of all. By Saturday the cold will moderate, the wind will tire and gradually die to a calm or become weakly variable. The four day supremacy of the anticyclone will be over. But, mind you, there are a dozen variations of this routine. I am only suggesting a usual one.

If after blowing two or three days from the west the wind shifts to the southwest and south, you may know that the central cold area is passing south of you and that its intensity will not be great. While these anticyclones that float down and to the right of their normal path linger longer, they are never so severely cold, nor, alas, so uniformly clear as the others. It is a profound law of anticyclones and even more particularly of cyclones, that if they deviate to the right they weaken, if they are pushed by an obstacle to the left they increase greatly in intensity.

Occasionally the central portion of an anticyclone passes over your locality. Then the wind will fall. The frost will be keen and the cold will be notably dry and invigorating. In summer although the sunlight may be powerfully bright and the heat great, yet the air will have a buoyant effect, the body a resilience. And the nights will cool swiftly. Soon after the center passes from the locality a wind will spring up from the east with rapidly rising temperature and increased humidity.

The coldest part of the anticyclone is not, as one would suppose, at the center, but in advance of it; and its authority, like a schoolmaster's, is rapidly dissipated after its back is turned upon a place.

The intensity of an anticyclone is measured by its wind velocity and by the degree of cold obtaining under its influence. But the greatest cold occurs rarely in conjunction with the greatest velocity of the wind. The calms that occur at sunrise enable radiation to take an extra spurt which pushes the mercury lower by a degree or so than happens when the wind is blowing. But, windy or calm, the period about sunrise is normally the coldest of the day, even extending in midwinter for as much as half an hour after sunrise, so slow are the feeble rays at restoring the balance of loss and gain of heat.

The greatest falls occur at the advent of the cold wave, no matter whether it arrives at ten in the morning or at midnight. If the temperature starts to decline gradually during the day, a further and decided fall may be expected at nightfall if the sky is clear. And if the temperature rises gradually during the night the normal processes are being displaced and a change from fair to foul is a surety. In summer the hottest time of day is not at noon, any more than the coldest part of the winter day was at midnight, for the reason that the sun can pour in its heat faster than the earth can radiate it, and the hour for the maximum temperature is pushed as far along toward evening as four or five or even six o'clock.

The average anticyclone continues its influence for clearness for about four days. Some, however, hurry the whole thing through in two. Others are interrupted by a more vigorous cyclone and are put to rout. Others are held up by an inherent weakness and are forced to mark time over one locality until strengthened or dissipated. And a few great ones hold sway over the country for a week. These choose the north-center of the country in which to locate. There they pile up the cold air until its very weight causes it to move majestically on. Its skirts sweep the Gulf coast where they are a bit bedraggled by invading cyclones. It gives the New Englanders a fortnight of nipping, brisk days and the mercury in Minnesota and the Dakotas does not emerge above zero. Once, in Montana, one of these refrigerating systems established the record of sixty-three degrees below zero. But in Siberia where the immense extent of the land surface collaborates with a prolonged night, an anticyclone built up an area of superior chilliness that left a world's record of ninety-one below.

In summer a succession of these highs causes the frequent droughts of weeks which harass the West and New England. The air becomes so dry that it parches and then shrivels the green leaves. Any little cyclones that, under ordinary conditions, would suck in moist air from the Gulf and relieve the situation with a rain are dried out and frustrated by the unclouded sun. It requires a cyclone of great depth to overthrow the supremacy of these summer anticyclones.

While the anticyclone furnishes fair weather the sky is not necessarily or even usually free from clouds under its influence. In summer the evaporation during the long days overloads the air for the time being. Normally about eleven in the morning little balls and patches of white clouds dot the blue. These increase in number and size until about three in the afternoon when they will have grown little black bellies and fluffy white tops. By five they will have dwindled and by eight entirely vanished. These heaped clouds, known as cumulus, are a guarantee of a normal atmosphere and continued fair weather. They mean that currents of warm, moist air have risen until they have struck a level so cool as to cause them to condense part of their moisture. This condensation sinks until it enters a warmer stratum and the cloud is dissipated. The total movement is a reasonable exchange that preserves the equilibrium of the air, very much as a person bends one way and then another to maintain his balance.

In winter there is not such an opportunity offered and the few clouds that form because of the daily variation in temperature are flatter and are called stratus clouds. Sometimes these stratus clouds may cover the sky at midday, but in thin platings and not leadenly. In winter as in summer they tend to disappear toward evening. They are often accompanied by an unpleasant wind, but rarely by the snow flurry which is the "April shower" of the winter months.

But when the snow flurry does come there is no better sign for the woodsman of coming cold; it never fails. The morning will have begun brilliantly, but soon great summery puffs of cloud form and increase and darken on their under sides. Their tops are vague and wear a veil. It is the snow. The reason is simple. The coming anticyclone strikes the upper air before it hits the earth's surface. The sudden cold causes rapid condensation. Hence the flurries. But the anticyclone is an agent of dryness, hence their short duration. Sometimes the veil of snow does not reach the earth. Sometimes it blots out everything in a spirited squall. But it never lasts long, except in the northwest states. And it is invariably followed by a period of colder weather.

In summer local evaporation may be so long-continued or so vigorous that the cumulus clouds cannot hold all their moisture content when cooled. A shower is the result, usually a trifling one and mostly without thunder. The great thunderstorms are always in connection with the passing of a cyclone. The small heat thunderstorms are only the indulgences of a spell of fair weather. These tiny showers are daily and sometimes hourly accompaniments of clear weather in the mountains. The air warms rapidly in the valleys and is speedily cooled on rushing up a mountain side and a threat and a sprinkle are the result. When a performance of this sort is going on nobody need fear unpleasant weather of long duration.

Another pledge of a clear day that does not appear too credible on the face of it is the morning fog in summer. In winter it is a different matter. In August and September particularly the rapidly lengthening nights allow so much heat to evaporate that the surplus moisture in the air is condensed to the depth of several hundred feet. By ten o'clock the sun has eaten into this lowest stratum, heated it and yet begins to decline in power before the balance swings the other way, so that a cloudless day often follows a fog in those months. About three mornings of fog, however, are enough to discourage the sun and a rain follows. Of course this is because the anticyclone with its special properties has been losing power.

When these conditions of clear nights with no wind follow the first two or three windy days of the anticyclone, particularly in autumn and spring, frost results. In winter the chances that a fog will be dissipated are rather slim. But if it shows a tendency to rise all may yet be well.

The fine-spun lines of the cirrus proper drag this veil of whitish cloud over the sky. The sun sometimes is surrounded by a colored halo due to the refraction of the light by the ice crystals. But more often it vanishes behind the veil. The mottled clouds below the veil show that a rather rapid condensation of the moisture in the air is taking place. This sky is distinctly threatening, although the direction and force of the wind will more accurately foretell the severity of the coming storm. With this sky expect rain or snow within 12 hours.]

An excellent sign of clear weather is this fact of the morning mist rising from ravines in the mountains. And even if you haven't any mountain ravines at command the altitude of clouds can be observed. It is safer to have them lessen in number rather than increase, scatter rather than combine. The higher the clouds the finer the weather. And if the sky through the rifts is a clear untarnished blue the prospects of settled weather are much better than with fewer clouds and a milky blue sky beyond.

After the direction of the wind and the shapes of the clouds the colors of the sky are a great help in the reading of the morrow's promise. And the best time to read this promise is in the morning or evening when the half lights emphasize the coloring.

Soon after the close observation of cloud colors has commenced the amazing discovery is made that the same color at sunrise means exactly the reverse of its meaning at sunset.

"Sky red in the morning A sailor's sure warning, Sky red at night A sailor's delight."

Christ seized upon this phenomenon to throw confusion into the Pharisees and Sadducees when they asked that He would show them a sign from Heaven. As Matthew reports it:--"He answered and said unto them, When it is evening ye say, It will be fair weather for the sky is red. And in the morning It will be foul weather to-day for the sky is red and lowering. O ye hypocrites, ye can discern the face of the sky; but ye cannot discern the signs of the times."

The reasons for this contradictory evidence of color are not nearly so obvious as the fact itself. Taking the scientist's word for it need not stretch one's credulity overmuch if he can be followed step by step. He says that sunlight is white light, and white is the sublime combination of every color. If no atmosphere existed about us the light would all come through, leaving the sky black. The atmosphere, however, which is full of dust and water particles, breaks up these rays, these white sheaves of light, into their various colors. The longest vibrations, which are the red, and the shortest, which are the violet, get by and the rest are turned back, mixing up into the color which we call our blue sky.

If the dust and water particles grow so large and numerous as to divert more of the short rays than usual we get a redder glow than usual. This is most noticeable when the sun and clouds are near the horizon for the air through which they appear is nearer the earth and consequently dirtier. If these water globules mass together so as to reflect all the rays alike the result is a whitish appearance. That is why a fog bank, composed of tiny droplets, each reflecting with all its might, can make the sky a dull and uniform gray.

As evening approaches the temperature of the normal day lowers. As the temperature lowers it is the tendency of the moisture in the air to condense about the little dust particles in the air. And as these particles increase in size their tendency is to reflect more and more of the waning rays of light. Therefore if the sky is gray in the evening it means that the atmosphere already contains a good deal of condensed moisture. If the cooling should go on through the night, as it normally would, condensation would continue with rain as the likely result.

If, on the other hand, after the evening's cooling has progressed and yet the colors near the horizon are prevailingly red it means that there is so little moisture in the atmosphere that the further increase due to the night's condensation will not be sufficient to cause rain. Hence the natural delight of the sailor.

A gray morning sky implies an atmosphere full of water precisely as an evening gray does. The difference lies in the ensuing process. By morning the temperature has reached its lowest point and if this has not been sufficient to cause cooling to the rainpoint the chance for rain will be continually lessened by the growing heat of the rising sun. The gray, therefore, is the normal indication of a clear cool night which has permitted radiation and therefore condensation to this degree. It is for this reason that we have the heavy fogs of August and September followed by cloudless days.

A red morning sky shows, like the red evening sky, that condensation has not taken place to any extent. But this is abnormal for a clear night causes condensation. The red therefore means that a layer of heavy moist air above the surface levels has prevented the normal radiation. Hence when the day's evaporation adds more moisture to that already at the higher levels the total humidity is likely to increase beyond the dewpoint with the resultant rain.

These two color auguries are among the most reliable of all the weather signs. Unfortunately the sunrises are scarcely ever on hand to be examined except by milkmen. But a careful scrutiny of the sunset will make one proficient in shades. In summer when the sun burns round and clear-cut and red on the rim of the horizon the air contains much dust and smoke, the accompaniment of dry weather. And as dry weather has a way of perpetuating itself such a sun makes dry and continued weather a safe prophecy. In winter the same red and flaming sun setting brilliant as new minted gold is a sure indication of clear and cold weather. In all seasons the light tints of the evening sky mean the atmosphere at its best. A golden sunset, a light breeze from the west, a glowing horizon as the sun goes down, slow fading colors all constitute a hundred to one bet for continued fair weather. The sunset colors that are surely followed by storm will be discussed in the next chapter.

The sky is too little regarded. Architects that do not consider the sky are behind in their calling. Maxfield Parrish has made himself famous by allying himself with its seas of color. The hunter can read it and learn whether he may sleep dry without his tent. Only we who shut ourselves within rooms and behind newspapers forget that there is a sky--until it falls and we are taken to a sanitarium.

From the night itself much may be discovered about the continuance of fair weather. A sky well sown with stars is a good sign. If only a few stars are visible the clear spell is about over. Stars twinkle because of abrupt variation in the temperature of the air strata. If the wind is from the west cold and clear will result no matter how much may twinkle twinkle little star. But if he twinkle with the wind from the south or east the cloud will soon fly. That is the way with these weather signs. One sign does not make a prophecy. It is the combination that has strength and reliability. Furthermore the eye must be trained by many comparisons.

Of all the conditions that make night forecasting easy the later evenings of the moon are the best. The moon furnishes just the proper amount of illumination to betray the air conditions. If she swims clear and triumphant well and good. If she rides bright while dark bellying clouds sweep over her in summer, inconsequential showers may follow. But if she disappears by faint degrees behind a thin but close knit curtain of cloud the clear weather is being definitely concluded.

A great many changes in the weather take place after three in the morning. Most campers are accustomed to waking anyway once or twice to replenish the fire, and a glance at the stars will show the sleepiest what changes are occurring in the eternal panorama. A man may have gone to bed in security to get up in a snowstorm, whereas a survey of the skies at three would have noted the coming change. The habit of waking in the dead of night,--which isn't really so dead after all,--is not an unpleasant one. Its compensations are set forth in a beautiful and vivid chapter of Stewart Edward White's "The Forest." Every camper knows them, and this added mastery that a knowledge of the skies gives him lends a sense of power, which lasts until the unexpected happens.

For the unexpected happens to the best regulated of all forecasters, the Government. Equipped with every instrument and with an army whose business is nothing else than to hunt down storms and warn the public, the Weather Bureau is still surprised fifteen times out of a hundred by unforeseeable changes in atmospheric pressure. It is scarcely likely then that amateurs without flawless barometers and without reports of the current weather in three hundred places could hope to foretell with complete accuracy. But there is a place for the amateur, aside from his own personal gratification and profit. The Weather Bureau within the limits of the present appropriation cannot expect to predict for every village and borough. That the amateur may do and with as great accuracy for the few hours immediately in advance.

The Weather Bureau may predict with this large percentage of accuracy--85%--for forty-eight hours in advance because its scope is country wide. It may even forecast in a general way for seven days and still maintain a considerable advantage over almanac guesswork. But the man who is relying upon local signs is limited to ten or at most twelve hours. Of course he may guess beyond that but it is only a guess. The work that the Bureau does and that he may do within his limits is not guesswork. Meteorology is an exact science, and forecasting is an art. Both may be studied now in classes under professors with degrees in the same way that any other science and art may be studied. The old sort of weather wisdom which was a startling compound of wisdom, superstition, and inanity has passed away, or is passing away as rational weather talk spreads.

These limits of the layman--ten hours with no instruments--are further defined by his locality. In mountainous country changes come more quickly than in level localities, in winter than in summer, so that one's prophetic time-limit is shortened.

While the best indications of the clear day are the great fundamental ones, there are many little signs that bolster up one's confidence in one's own predictions. The lessened humidity coincident with clear weather is responsible often for many little household prognostics. Salt is dry. The windows (of your summer cottage) do not stick. The children are less restless. Smoke ascends, or if the wind is blowing is not flattened to the ground. Flies are merely insects, for the time being, and not the devil. Swallows and the other birds that eat insects fly high because that is where the insects are. Spiders do not hesitate to make their webs on the lawn. They welcome dew but distrust rain. Cow and sheep feed quietly, rarely calling to one another as they do before a storm. In short the general aspect of these is normal and therefore remains unnoticed.

But all these household prognostics may be advertising the most placid weather while only twelve hours away and coming at sixty miles an hour may be the severest storm of the season. The Weather Bureau with its maps and barometers follows its every movement. The man in the woods whose comfort in summer and whose life in winter may depend upon his preparedness for the approaching storm does well to read its warnings and know its laws.

CHAPTER III

THE STORM CYCLE

Doubtless those who hope for a Hereafter of unmitigated ease and song, desire, on this earth, one long, sweet anticyclone. But theirs, in most of the United States, is disappointment. With an irregularity that seems perversely regular at times our fair weather is interrupted by a storm which in turn gives way to some more fair weather or another storm,--there is no telling which very long in advance. And that is why American weather ranks high among our speculative interests.

To emphasize this irregularity a seemingly regular succession of events may be noticed. It will cloud up, let's say, on a Sunday, rain on Monday and Tuesday, clear on Wednesday, staying clear until Sunday when it will cloud up for the repeat. During this past season it rained on a dozen washdays in succession. The newspapers grew jocular about it. And very often one notices two or three rainy Sundays in a row. By actual observation this year we enjoyed fifteen clear Thursdays in succession in a normal spring.

The weather gets into a rut. And if the anticyclones and cyclones were all of the same intensity it is conceivable that the rainy Sundays might go on until the Day of Rest was changed by Statute. But the intensities of the whirls differ. Before long an anticyclone feebler than ordinary is overtaken by a cyclone and annihilated. Or one stronger than the average may dominate the situation for several days. Or the great body of cold air in winter over the interior of Canada may send a succession of moderate antis across our country making a barrier of dry cold air through which the lurking cyclone can not push.

Mostly, however, three days of anticyclonic influence and three days of cyclonic influence with one day in between for rest, the transition period, make up a normal week of it. Let the American farmer thank his stars (and clouds) for that. For no other regions of the earth are so consistently watered and sunned all the year round as the great expanse of the North American Continent.

The cyclone is that activity of the atmosphere which prevents us from suffering from an eternal drought. The cyclone is an accumulation of air which has become warmer than the air about it. This area of air usually has a central portion that is warmest of all. Since warmth expands, this air grows lighter and rises. Nature, steadfast in her grudge against a vacuum, causes the surrounding air to rush it. Since these contending currents cannot all occupy the central area at once they fall into a vast ascending spiral that spins faster and faster as it approaches the center. Imagine an inverted whirlpool. It is a replica on a much smaller scale of the great polar influx, except that the latter has a descending motion.

The cyclone thus is tails to the anticyclone's heads, the reverse of the coin. Where the anti's air was cool and dry the cyclone's is warm and moist. The anti had a downward tendency and a motion, in our hemisphere, flowing outward from the apex in generous curves in the direction of clock hands. The cyclone has an upward tendency, flowing inward to the core contrariwise to clock hands.

From these two great actions and reactions come all the varieties of our weather. To understand the procession of the cyclones and anticyclones across our plateaus, our mountains, our plains, and our eastern highland is to know why, and often when, it will be clear or not. To mentally visualize the splendid sweep of the elements on their transcontinental run is to glimpse grandeur in the order of things which will go far to offset the petty annoyances of fog or sleet. Ignorance may be bliss, but knowledge is preparedness.

The anticyclone suggests a pyramid of cold, dry air. The cyclone suggests a shallow circular tank in leisurely whirl. But all comparisons are misleading and a caution is needed right here. For a storm is not a watering cart driven across our united skies by Jupiter Tonans Pluvius. It is NOT a receptacle from which rain drips until the supply is exhausted. A cyclone is a much more delicate operation than that. It is a process. It can renew itself and become a driving rain storm after it had all the appearance of being a sucked orange for a thousand miles.

Suppose that our cyclone, this organization of warm, moist air with its curving winds, enters the state of Washington on a Wednesday, from the North Pacific. As early as the Monday afternoon before the wind throughout all that section of the country would have shifted out of the west and have started to blow in some easterly direction,--northeast in British Columbia and southeast in lower Idaho. But since these winds are blowing from the interior they are dry, and consequently rain does not fall much before the storm center is near, that is on the Wednesday. If the storm center passes north of Tacoma the winds, shifting by south and southwest, bring in the ocean moisture and heavy rain commences which continues until the rising barometer and westerly winds indicate the approach of another anticyclone. So much for western Washington.

As the cyclone passes eastward it mounts the Cascades and its temperature is lowered, its moisture is squeezed out, and it stalks over Montana, the mere ghost of its former self, as far as energy and rainfall are concerned. To be sure it preserves its essential characteristics of relative warmth, and inwhirling winds. But let it continue. As its influence begins to be felt over Wisconsin and the Lake region the moister air is sucked into the whirl and rain, evaporated from Superior, falls on Minnesota. The east winds are the humid ones now, the west ones the dry. Eastward the center moves, over Indiana, Ohio, New York, the rainfall steadily increasing as the ocean reservoirs are tapped.

The first time you tell a New Englander that his easterly storms come from the west you are in danger, unless he be a child, for it is to the children that one may safely appeal. Indeed it is the increasing number of children who are learning these fundamental weather facts in the public schools that the Weather Bureau relies upon for a more intelligent support in the next generation. They teach their parents. These latter find it difficult to believe, however, that the storms which hurl the fishing fleets upon the coast in a blinding northeaster have not originated far out at sea, but have come across the continent. For the safe handling of boats knowledge of the rotary motion of storms is necessary that one may be able to tell by the direction of the winds and the way they are shifting where lies the center of the storm and its greatest intensity.

In Tacoma when the wind shifted by way of southeast, south, and southwest that was proof that the storm center was passing north of the city. Likewise if in New York the winds shift by way of northeast, north, and northwest the storm center is passing south of that city. As it drifts out to sea it is gradually dissipated by the changing influences on the North Atlantic. Very few of our storms ever reach Europe, although some have been traced to Siberia.

The Government has put its sleuths on the track of every storm that has crossed the United States in the last thirty years. These weather detectives with a thousand eyes have made diagrams of their actions, mapped their courses, computed their speeds, and if we don't know where all our discarded storms go to, we at least know where most of them came from and how they acted when with us.

About a hundred and ten areas of low-pressure affect the country during the normal year. Of these all but seven, speaking in averages, come from the West so that the Boston mechanic who will not believe that the nor'easter comes via the Mississippi Valley is right about 7/110 of the time. But even that small fraction is no exception to the general law, because those seven storms are not born in Newfoundland but in our East Gulf States. They come up the Coast, and the wind blows from the northeast and north into their centers while they are still on the Carolina coast. The great hurricanes which are cradled in the tropics and march westward under the influence of the trades are genuine exceptions to the general westward rule, although they always eventually turn toward the east. They will be given the prominence they demand later, since the eastbound schedule must not be sidetracked now.

The wispy edges of the cloud at the brightest part are cirrus, the fleecy cloud at the extreme top is a thin alto-cumulus, and the dark base of the sky is stratus. But this stratus is too high for that classification and so they call it alto-stratus. This sky shows that the temperatures are moderate, a cold sky being much better packed, and a warm one fluffier. The fact that a veil of cirrus has not preceded the heavier clouds argues that the coming storm will not be of much consequence. This sort of cloud bank arising after a period of cold weather is the best possible prediction of a thaw. Slight rain might follow within a few hours.]

Three cyclones a year form over the lower Ohio River basin. On account of their origin over land instead of over water they rarely acquire much energy. Once in a decade such depressions deepen rapidly. It was one of these Ohio River storms that increased greatly in energy while moving from West Virginia to the Jersey Coast that gave Philadelphia her Christmas Blizzard, a surprise to her citizens and to the Weather Bureau, for most of the snow fell with the mercury above freezing. The flare-back which gave Taft his big inaugural snowstorm is another example of the way a depression may deepen on approaching the coast. Until the upper atmosphere is as well understood and watched as the lower, or until instruments are perfected whereby the weather conditions can be made self-announcing such surprises are absolutely unavoidable. Under conditions that warrant any suspicion of sudden developments the Bureau at Washington is careful to order extra observations in the areas likely to be affected, but no surface observations can quite suffice.

Fifteen storms a year originate over the west Gulf States, or, drifting in from the Pacific over Arizona and New Mexico begin to acquire energy in Texas. Twelve are set up over the Colorado mountains. These usually dip down into Texas before starting their drive toward the northeast. After both these sets of storms get under way they strike resolutely for the same locality,--the St. Lawrence Valley. The conformation of the St. Lawrence region provides an irresistible attraction for American storms. Occasionally a very strong anticyclone holds that territory and pushes the cyclone off the coast at Hatteras or even makes them drift across the country to Florida. But such occasions are exceptional. Give the ordinary cyclone its head, and, ten to one, you will find it on the way to the St. Lawrence. The inhabitants will confirm this statement, I am sure. They do not feel discriminated against in the matter of weather. They get nearly everything that is going. Since they have to accommodate from seventy to eighty cyclones in fifty-two weeks they have very little time to brood over any one variety of weather. With the optimism of that section of the country they say, "If you don't like our weather, wait a minute."

Ten storms a year originate over the Rocky Mountain Plateau, north of Colorado. About twenty cross over from the Canadian Provinces of Alberta and British Columbia. And all our other storms, about forty each year, enter our country from the North Pacific by way of Washington and Oregon. Many of these drift across the northern tier of states without any great display of energy, at least before they reach the Lake region. But the majority loop down somewhat into the middle west as far south as Kansas, and then make their turn toward the inevitable St. Lawrence. They usually require four days to make the trip from coast to coast by this route, as also by the more direct northern route, because on that they travel more at leisure. But the storms from Texas, whose energy is greatest because of greater heat and moisture, occasionally speed from Oklahoma to New York in thirty-six hours.

In summer all speeds are reduced. This is because the disparities in temperature are less. In winter where greater extremes of temperature are brought into conjunction the processes of the storm are all more violent. And it is a bit disheartening to know that a storm is aggravated to even greater endeavors by its own exertions. Its energy provides the conditions to stimulate greater energy, and, like a fire, it increases as it goes. If it did not run out of the zone which nourished it and proceed into another zone where conditions were distinctly discouraging the limits of the storm would be much extended, and vast territories would be devastated by the self-propelling combination of wind and water.

To the generality of us the word storm means rain. To the scientist it means wind. In reality the cyclone is rare that crosses our country without causing rain somewhere along its track. The curiosity of the Weather Bureau to find out the paths of the storm centers is abundantly justified because it is along these paths that the heaviest rainfall and the severest winds occur. But whether or not there is precipitation on the path of the cyclone it is rated as a storm if there is a lowering of pressure and consequent wind-shift.

The storm centers are not always well-defined, and quite often the circulation of the wind about them is not complete. Such cyclones never amount to much, although there is always the possibility of their closing in and developing a complete circulation with the attendant increase of energy. The incomplete cyclones over the desert and plateau regions are lame affairs, lacking interest and advancing timidly if at all. But once let them drift into a locality where they can be supplied with moist air, they pick up energy, keep a definite course, and advance with increasing speed.

Very often the center will split up, the circulation perfecting itself around both centers of depression. One of these will likely be over Minnesota and the other over Texas and the organization will steam-roller the states to the east in the manner of a gigantic dumb-bell. This formation is more likely to have been caused by the two centers appearing simultaneously than by a split in an original center. The weather reports call this fashion of storm a trough of low pressure. The southern center is the one that develops the more energy on its turn to the northeast. If the two centers should unite on reaching the northeast a very heavy precipitation is the invariable result.

All cyclones have much greater length than breadth. They frequently stretch from unknown latitudes in Canada into unrecorded distances into the Gulf, while on the other hand it is a very large storm that rains simultaneously upon the Mississippi and the Atlantic. Behind a cyclone of pronounced energy a second whirl, called a secondary depression, often develops, in which case the period of wet weather is prolonged. Also, more rarely, an offshoot forms ahead of the main depression.

A sluggish, sulky cyclone either in winter or summer provides more opportunity to humanity for self-discipline than almost any other feature of our national environment. In winter when the depression slows up it settles down upon one community in the guise of fog, and stays by the locality until an anticyclone blows in and noses it out. Fog is aggravation, but a hot wave is suffering and the hot wave is caused by a depression weak in character but generous in dimensions getting held up on the northern half of our country. By its nature it attracts the air from all sides, and being in the north, the direction of the wind over most of the country would be southerly. Air from the west and north has a downward tendency, but south and east winds are surface currents. Consequently these winds, blowing over leagues of heated soil, become dry and parching. If the depression lingers long the entire country to the east, south, and west soon suffers from superheated air. At last the very intensity of the heat defeats itself and the reaction to cooler is effected dramatically through a thunderstorm.

The well-developed cyclone in winter causes what we all know as a three days' rain, although continuous precipitation rarely lasts over ten hours. The rest of the time is occupied by general cloudiness with occasional sprinkles and a final downpour as the wind shifts to the west and the anticyclone nears. In summer the depressions, being shallower, rarely cause continuous cloudiness for three days, although their influence often lasts as long as that in the guise of a series of thunderstorms. The line of storms extends several hundred miles, bombarding all the towns from Albany to Richmond. These thunderstorms sometimes achieve in an hour or two even greater results than their winter relatives can accomplish in three days in the matter of rainfall, wind velocity, and general destructiveness. Our wettest months are July and August and not December and January.

The freedom of the wind has been the subject of much poetic and prosaic license. As a matter of fact the wind is the veriest slave of all the elements. It is harried about from cyclone to anticyclone, wound up in tornadoes, directed hither and thither by changing temperatures. It blows, not where it listeth, but where it has to. And circuitously at that. For once the path of duty is not straight. That is another fact that the Boston mechanic would have been slow to accept,--that the wind blows in curves. A little consideration, however, of the fact that the wind is perpetually unwinding in great curves from the anticyclone and winding up on the cyclone will show that nowhere can it be blowing in a perfectly straight line.

Thus it becomes the surest indication that a cyclone is to the west of one if the wind blows from an easterly point. The storm is bound to move toward the east, therefore the rapidity with which the clouds move and thicken will signify when the area of precipitation will reach the observer. The cycle of the storm is normally this: After a cloudless and windless night a light air springs up from a little north of east. At the same time strands of thin wavy clouds appear, very high up. They may be seen to be moving from the southwest or northwest. Their velocity is great. Their name is cirrus, and they are called mares' tails by the sailors. They are followed by several hours of clear skies, usually; but if the storm is smaller and close at hand there is no clear interval.

Before the larger storms these cirrus clouds are sent up as storm signals twenty-four and even forty-eight hours in advance. The day that intervenes is very clear, the air feels softer, the temperature is higher. In midafternoon more cirrus appears, and as condensation follows the quick cooling the silky lines increase in number. Beneath them a thicker formation, known as cirro-stratus, forms a dense bank in the west and southwest. The sun sets in a gray obscurity. If there is a moon it fades by degrees behind the veil of alto-stratus, and the halo which first was seen wide enough to enclose several stars narrows until it chokes the moon in its ever-thickening cocoon of vapor.

There is no value whatever in the old superstition that the number of stars within the halo foretells the number of days that it will rain or snow. The same halo that encloses three stars at eight o'clock may have narrowed down to one by midnight, or none at all, so that the prophetic circle is bound in the very nature of its increase to contradict itself. The presence of a halo is a pretty sure sign of some precipitation within twenty-four or thirty hours. It fails about thirteen times in a hundred. If the halo is observed around the sun it is an even surer sign, failing only seven times the hundred.

During the time of cloud-increase the wind will probably lull before a snow, so that the hour or so before precipitation begins is one of intense brooding calm. Or if there is no calm the wind, now easterly, will be very gentle. Soon after the precipitation begins the wind will begin to freshen and will continue to increase in velocity until the center of the storm is close to the locality. This will require about eight hours for the average storm. As storms vary an average is a very misleading thing and the best way to judge of the length and severity of the storm is by watching the wind. If it increases gradually the storm will be of long duration. If the wind rises fitfully and swiftly it will not likely be long but may be severe. If the wind reaches any considerable velocity before the rain or snow begins the storm is sure to be short and severe.

The color and formation of the clouds will tell when the precipitation is about to begin. In summer, no matter how striking and black are the shapes and shadows of the clouds, rain will not fall until a gray patch, a uniform veil called nimbus is seen. In the little showers of April this patch of unicolored cloud is there, as well as behind the great arch of the onrushing thunderstorm. In winter raindrops are smaller and the tendency of the clouds is to appear a dull, uniform gray at all times. But the careful observer can detect a difference between the nature of the clouds several hours before precipitation and their color immediately before.

When snow is about to fall no seams are visible. An impenetrable film obscures all the joints. From such a sky as this snow is sure to fall. But if seams are visible, if parts of the skyscape are darker than others, then, no matter whether the temperature on the ground is below freezing a rain storm will ensue. Very often these winter rains begin in snow or sleet, but the clouds register the moment when the change from snow to rain is to be made. The presence of swift-flying low clouds from the east is a certain sign that the change to a temperature above freezing has been effected in the upper strata of the atmosphere. This variety of cloud is called scud, and accompanies rain and wind rather than foretelling it long in advance.

If the storm is approaching from the southwest the precipitation begins near the coast about twelve hours after the cirrus clouds commence to thicken and about twenty-four after they were first seen. In some localities as much as thirty-six and even forty-eight hours are sometimes required for the east wind to bring the humidity to the dew-point. Just a little observation will enable one to gauge the ordinary length of time required to bring things to the rain-pitch in one's own country. Of course no two storms in succession make the trip under the same auspices and with the same speed. The sign of the Universe should be a pendulum. One period of cyclone, anticyclone, cyclone will traverse the country rapidly. Then there will be a halt all along the line, and the next series,--anticyclone, cyclone, anticyclone, will take three days longer to make the crossing. Otherwise our weather would have a deadening regularity.

On an average our storms cross the country at the rate of about six hundred miles a day. This is the average. Some delay, linger, and wait for days over one locality. Others do a thousand miles in the twenty-four hours. They thicken up enough to cause rain from two hundred to six hundred miles in advance of their centers. It stops raining not long after the actual center has passed.

But for picnic purposes the storm is far from being over. For even though continuous raining has stopped the low pressure still induces a degenerate sort of precipitation called showers, or oftener mist for another twelve hours (usually in winter). Then as the cooling influence of the anticyclone approaches the rain recommences. This time it is not for long, however, and is followed by permanent clearing, the wind shifting into the west. Sometimes the change to blue sky is abrupt. But if the subsequent anticyclone is not very well defined, cloudy conditions may linger for a couple of days. Such clouds are usually much broken and show white at the edges and never cause more than a chilly feeling.

This attempt to outline the customary cycle of the storm,--clear sky, cirrus cloud, wind-shift to the east, the denser cirro-stratus, the pavement-like stratus, the woolly nimbus, the first continuous hours of rain, the misty interval, the windshift to the west, the final shower, and breaking cloud, the all-blue sky--this storm-schedule is always subject to change. But the fundamentals are there in disguise every time. They only have to be looked for and there is some satisfaction in penetrating the disguise.

When a storm comes up the Atlantic Coast, as happens a few times a winter, the process is shortened, because the effects of the larger easterly quadrants are felt only at sea. The most prominent recent illustration of this type of storm was the severe snowstorm that swept the coast states from Carolina to Maine the Saturday before Easter, 1915. Its calendar read as follows: Friday, 8 P. M., cirrus clouds thickening into cirro-stratus. Midnight, stars faintly visible, wind from northeast, 12 miles an hour. Sunrise, stratus clouds, wind rising in gusts at Philadelphia to 30 miles; 8 A. M., rapid consolidation of clouds with snow shortly after, although the temperature at the surface of the earth was as high as seven degrees above the freezing point. This rapidly dropped to freezing. Flakes were irregular in size. Until one o'clock in the afternoon the snow thickened with gusts of wind up to forty miles. Snowfall for five hours was 14 inches, an unprecedented fall for this locality.

Then the storm waned for five hours more, 5 inches more of snow falling. Precipitation practically ceased at 6 P. M. By sunrise on Sunday the skies were free of clouds and the wind blew gently from the northwest.

Occasionally a high pressure area out at sea and beyond the ken of the Weather Bureau causes one of these coast storms to curve inward to the surprise of everybody. Occasionally, too, the transcontinental storms are driven north or south of their accustomed paths. While the divergence may be slight, it causes a margin of variance from the accuracy of the Bureau's report. Then arises a second storm,--one of indignation--from all the people on one side of the strip who carried umbrellas to no purpose, and from the others,--who didn't.

This pushing aside of the cyclone is caused by pressure variation that only hourly reports from many localities could detect. Vast hidden influences shift the weights ever so little and the meteorological express is wrecked. But this happens, at most, fifteen times in a hundred, and remembering the unseen agencies to be coped with people are refraining more and more from the tart criticisms of former times, not in charity but in justice, although there is small tendency yet to forward eulogies to the Bureau in recognition of the eighty-five times it is right.

CHAPTER IV

SKY SIGNS FOR CAMPERS

The weather-wise, even more so than poets, are born. But that only goes to say that weather-wisdom can be fathered. For poetry and canoeing and the art of making fires, once the desire for these things is born, may be aided infinitely by observation and practice. Nobody can teach a man the smell of the wind. But the chap who feels nature beating under his heart can, by taking thought, add anything to his stature. So it is with those who are called weather-wise. An unconscious desire, a little conscious knowledge, a good deal of experimentation with the cycle of days, and you have a weatherman.

These chapters aim to put the little conscious knowledge into the hands of the people with the unconscious desire, so that when they take their week in the woods for the first time (and their month for the second time) they may enjoy the shifting scenery of the sky-ocean and, incidentally, a dry skin. For I take it that everybody will soon be camping. Maine and the Adirondacks have become a family barracks. It is Hudson Bay for bachelors. And over this expanse of woods and children the weather problem ranks with the domestic one. For naturally if a soaking would endanger his vacation the husband must not permit a rain,--unexpectedly. In all seriousness, it is of avail to know the skies if one is going into the wilds just as it is of avail to know what severed arteries demand, what woods burn well, and what mushrooms can be eaten, even though one can get along without knowing these things until perchance the artery is severed or the arched squall catches one far from shore.

At the very least, one grain of weather wisdom prevents a mush of discomfort. And if, fellow-camper, the following observations gathered on a thousand thoughtless walks do not tally (for the northeastern states) with yours, write me, so that in the end we may finally contrive together a completer handbook of our weather.

THE CLOUDS

Clouds are signposts on the highway of the winds. Every phase of the weather, except stark clearness, is commented upon by a cloud of some sort. When danger is close they thicken. When it passes they disappear. The aviators of the future will be cloud-wary. He who flies must read or never fly again.

The cirrus cloud is always the first to appear in the series that leads up to the storm. It looks like the tail of Pegasus and for it the old forecasters in their forecastles made a special proverb.

"Mackerel scales and mares' tails Make lofty ships carry low sails."

These white plumes and scrolls which are in reality glistening ice-breath, fly at the height of five, six, seven, and even eight miles. And as a sign of coming storm they are about as infallible as anything may be in this erratic world. They were born in the cradle of a storm. The storm center was breathing warmed air upward to great heights, and although the disc of the storm itself was only two or three miles deep, its nucleus, crater-like, shot warm columns twice as far. With just enough moisture content to make a showing against the blue these streamers flowed to the eastward. At those dizzy heights the prevailing westerlies are in full force, blowing from eighty to two hundred miles an hour night after night and day after day. These westerlies caught the storm exhalations, the streamers, and hurled them eastward at greater speed than the main body of the storm. And that is the reason that we see these cirrus clouds always eight, mostly twelve, often twenty-four and sometimes forty-eight hours before the storm is due.

Just a few strands of cirrus have little significance. They may be condensation from a local disturbance, or a back fling from a past storm. But if the procession of the cirri has some continuity and broadens to the western horizon it is a sign about eight times in ten that a cyclone is approaching. Occasionally the storm center is too far to the south or north to cause rains at your locality, but the cirri bank up on the horizon and their lacework covers the sky. If they appear to be moving toward the region of greatest cloudiness it is not a sign of precipitation. This condition is most apparent at Philadelphia when the storm center over Alabama or Mississippi floats out to sea by way of Florida without having the energy to turn north. Then the cirrus is seen thickly on our southern horizon. Looking closely one sees that the cirri are moving from the northwest, and are being drawn into the storm area instead of proceeding in advance of it.

Careful watching will sometimes enable one to tell whether the tails are increasing or decreasing in size. If they dissolve it means that the cyclone from which they were projected is losing strength because of new conditions. Cloudiness may follow but no precipitation of consequence. The plumy tails are expressive: pointing upward they mean that the upward currents are strong and rain will follow; pointing downward they mean that the cold dry upper currents have the greater weight and clear weather is likely. In summer the cirrus cloud formations are not such certain advance agents of rain because all depressions are weaker and less able to confront a well-intrenched drought. As the proverb goes, "all signs of rain fail in dry weather," and there is some truth in it.

The fine wavy cirrus clouds often increase in number, develop in texture until the blue sky has become veiled with a muslin-like layer of mist. This is the cirro-stratus, and is a development of the cirrus, but it does not fly so high. Its significance is of greater humidity and is the first real confirmation of the earlier promise of the cirri. Another form that the cirro stratus may assume is the mackerel sky,--clouds with the light and shade of the scales of a fish. If this formation is well-defined and following cirrus it is a fairly accurate storm indicator. It is not quite infallible, however, as the same forms may be assumed when the process is from wet to dry.

The old proverb, "Mackerel sky, soon wet or soon dry," expresses this uncertainty. If dry is to follow the scales will appreciably lessen in size and perhaps disappear. If the cirro-stratus or scaly clouds are followed by a conspicuous lowering it is only a question of a few hours until precipitation begins. The cirro-stratus at a lower level is called alto-stratus and this becomes heavy enough to obscure the sun.

The cloud process from stratus on is slow or rapid, depending upon the energy of the coming storm and the rate of its approach. In most cases the clouds darken, solidify, and become a uniform gray, no shadows thrown, no joints. Soon after the leaden hues are thus seamless the first snowflake falls. If it doesn't it is a sign that the process of condensation is halting: the storm will not be severe. Sometimes there is no precipitation after all this preparation, but under these circumstances the wind has not ventured much east of north. From the time that the snow starts the clouds have chance to tell little. Only by a process of relative lightening or darkening can the progress of the storm be followed and the wind, and not the clouds at all, is the factor to be watched; for occasionally the sun may shine through the tenuous snowclouds without presaging any genuine clearing so long as the wind is in the east.

But in summer the clouds become even more eloquent than the wind. The rain-cloud, called the nimbus, becomes different from the dull winter spectacle. In summer air becomes heated much more quickly and the warm currents pour up into the cold altitudes where they condense into the marvelous Mont Blancs (or ice-cream cones) of a summer afternoon. These piled masses of vapor are cumulus clouds, and if they don't overdo the matter are a sign of fair weather. They should appear as little cottony puffs about ten or eleven in the morning, increase slowly in size, rear their dazzling heads and then start to melt about four in the afternoon.

But perhaps the upward rush of warm, moist air has been so great in the morning that the afternoon cooling cannot dispose of it all without spilling. Then occurs a little shower,--the April sort. Often in our mountainous districts it showers every day for this reason. The great thunderstorms come for greater reasons: they are yoked to a low pressure area and represent the summer's brother to the winter's three-day storm.

Cumulus clouds are called fair weather clouds until their bellies swell and blacken and they begin to form a combination in restraint of sunlight. Even then it will not rain so much out of the blackness as out of the grayness behind it, and if there is no grayness chances are that you will escape a wetting. One can almost always measure the amount of rain that is imminent by the density of the curtain being let down from the rear of the cloud. If you can see the other clouds through it or the landscape the shower will be slight. If a gray curtain obscures everything behind it you had better pull your canoe out of the water and hide under it if time is less valuable than a dry skin. Such showers may be successive but rarely continuous.

Rain clouds have been observed within 230 yards of the ground. Very often it can be seen to rain from lofty clouds and the fringe of moisture apparently fail to reach the earth, because the condensation was licked up and totally absorbed on entering a stratum of warmer air. The reverse of this occurs on rare occasions;--condensation takes place so rapidly that a cloud does not have time to form, and rain comes from an apparently clear sky. This phenomenon has been witnessed oftenest in dry regions and never for very long or in great amounts, although a half hour of this sort of disembodied storm is on record.

If the cumulus clouds of the summer's afternoon do not decrease in size as evening approaches showers may be looked for during the night. And if the morning sky is full of these puffy little clouds the day's evaporation on adding to them will probably cause rain. A trained eye will distinguish between a stale and fresh appearance in cloud formation, the light, newly made, fresh clouds, like fresh bread, contain more moisture. If the clouds have much white about them they need not be feared as rain-bearers. Clouds are much higher in summer than in winter and the raindrops of warm air are larger than those of cool.

If cumulus clouds heap up to leeward, that is, to the north, or northwest on a south or southwest wind a heavy storm is sure to follow. This is notably so as regards the series of showers in connection with the passage of a low-pressure area. The wind will bear heavy showers from the south (in summer) for a whole morning and half the afternoon with intervals of brilliant sky and burning sun. Or perhaps the south wind will not produce showers, but all the time along the northwest horizon a bank of cloud grows blacker and approaches the zenith, flying in the face of the wind or tacking like a squadron against it. About the time that the lightning becomes noticeable and the thunder is heard the wind drops suddenly, veers into the west, and the face of things darkens with the onrush of the tempest.

Although no rain may have fallen while the wind was in the southern quarter yet that constituted the first half of the storm and the onslaught of rain and thunder the second. While the storm area moved from the west to the east the circulation of air about the center was vividly demonstrated by the south wind blowing into the depression, whose center was epitomized by the moment of calm before the charge of the plumed thunderheads from the northwest.

Most camping is done either in hilly or mountainous country where the movement of clouds is swifter and more changeable than over flat lands. There is one sign of great reliability: if the mountains put on their nightcaps the weather is changing for the wetter, and if clouds rise on the slopes of the hills or up ravines, or increase their height noticeably over the mountain-tops, the weather is changing for the dryer. In the mountains where abrupt cliffs toss the winds with all their moisture to heights that cool clouds form and condense rapidly and the weather changes quickly. But even in the mountains the big changes give plenty of warning.

Often clouds may be noticed moving in two or even three directions on different levels at once. The upper stratum will probably be cirrus from the west. Cumulus or stratus may be floating up from the south. A light drift of vapor called scud may fly on the surface easterly wind. Such a confused condition of wind circulation betokens an unsettled system of air pressures and as frequent collisions of the air bodies at varying temperatures are inevitable rains, probably heavy, will follow.

On clear days one will be surprised to see isolated clouds, usually the torn, thin sort, drifting across the sky from the east. A change will follow soon.

In winter black, hard clouds betoken a bleak wind.

Clear winter days several times a season show a brilliant blue sky filling with great cumulus clouds of dark blue, blurred at the top and gray at the base. They will sprinkle snow in smart, short flurries, and are ushering in a period of clear and much colder weather.

A sky full of white clouds and much light is a cheerful sign of continuing fair weather.

The softer the sky the milder the weather and the more gentle the wind. It is the dark gloomy blues that bring the wind. But do not mistake the woolly softness of the rolling clouds before a thunderstorm. A sudden and often violent gust follows. Tumbling clouds in any event should make one wary of venturing on water. Summer drownings would not be so numerous if the portent of the squall were heeded.

To this data might be added many singular cloud formations that are not observed often. The funnel shaped cloud of the tornado, the green shades of the hurricane cloud, the green sky of cold weather showing out between layers of steel blue, coppery tints that show before heavy storms sometimes, variations of color at sunset each of which has a meaning which practice in deciphering will make clear. But enough has been given to show sky-searchers how many are the tips of coming weather that may be read from a conglomeration of fog particles. Nobody with eyes should be caught unawares by day. The look of the sunset shadows forth much of the coming night. And throughout all this truth holds: the greater the coming storm the longer and clearer are the warnings given to the watchful.

THE WINDS

The wind is the ring-master of the clouds. It whistles and they obey. Therefore to be windwise is to be weatherwise, almost.

One can get a hold on the wind by learning to gauge its strength. Look at the trees or the smoke from your city chimneys and guess how fast it blows at eight o'clock in the morning, or eight at night. The weather report the next day will tell you how nearly you were right.

Beginning is easy; anybody can guess a calm. When the leaves are just moving lazily the Weather Bureau calls it a light or gentle breeze, moving from 2 to 5 miles an hour. A fresh breeze, from 6 to 15 miles will stir the twigs at first and finally swing the branches about. From 16 to 25 miles, a brisk wind, will cause white caps on the lakes, tossing the tops of the trees, but breaking only small twigs. Increasing from 26 to 40 miles it becomes a high wind that breaks branches on trees, wrecks signs in the towns, causes high waves at sea and roars like the ocean in heavy squalls through the woods. From 40 to 60 miles an hour makes a gale. Sailing craft are now in danger. The pressure at 50 miles an hour is 13 pounds to the square foot, having risen from three-quarters of an ounce at 3 miles. This pressure becomes 40 pounds per foot when the wind reaches a velocity of 90 miles.

At 60 trees are uprooted, chimneys may go, it is difficult to walk against, the noise becomes very great but rather inspires than frightens. As the gale increases from 60 to 80 (which velocity the Bureau rather weakly calls a storm wind), danger rapidly increases. Trees are prostrated, the uproar becomes terrifying, walking without aid is impossible, the great ocean liners are in danger, the sea becomes a whitened surface of driving spume that heaps up into piles of water thirty or more feet high, windows are blown in and frame houses cannot stand much greater velocities. Anything from 80 miles an hour up is well called a hurricane. Everything goes at 100. At Galveston the machine that registered the wind velocity blew away at 100.

They have better instruments now, and in many places velocities of over a hundred miles an hour have been recorded. As high as 186 miles was registered on the top of Mt. Washington, and in a single gust 110 at Montreal. The great hurricane winds are most felt at a few of the exposed places on our coasts. Cape Mendocino, on the Pacific, has 144 miles an hour to its credit in a January hurricane. But enough destruction is done at 90 miles. Fields are stripped of their crops, or leveled; houses are demolished unless they are specially built, like the New York sky-scrapers, to withstand much higher velocities. In the small whirling storms called tornadoes the wind is estimated to reach a velocity of 200 to 500 miles, and nothing but the cyclone cellar will shelter one from the fury of the elements when they are really unleashed.

The higher one goes the greater the velocity of the wind. On the top of Mt. Washington 100 miles is rather common for hours at a time and 150 is recorded now and then. That is only 6000 feet above Boston. If such a force struck Boston for a minute it would be blown en masse into the Bay.

Velocities on land are less than those at sea, because of the resulting friction from obstacles. Velocities in summer are lower (thunder gusts excepted) than in winter. Since the wind is caused by differences in atmospheric pressure, and that in turn by disparities in temperature, winter holds the palm for greater velocities because the wide whirl of a cyclone over the great plains may cause to mix air from Texas with a temperature of 60 degrees with air from Montana of 30 degrees below zero, while the summer temperatures in both states might easily be 80 degrees.

Throughout most of our land certain winds have always the same bearing upon the weather and this correspondence is roughly the same over most of the country. West winds, for instance, are an almost universal guarantee of clear weather. The Pacific Coast and western Florida are the exceptions.

Northwest winds bring clear skies and cool weather everywhere. In winter in the north plateau section heavy snows arrive in advance of the severe cold waves that come on these northwest gales.

North winds are the cold bearing ones. Clear skies prevail under their influence.

Northeast winds are cold, raw snow-bearing winds in winter and spring and bring chilly rains in midsummer.

East winds are the surest rainbringers of all for the eastern two-thirds of the country, and are soon followed by rain with a shift of wind over the other third. Their temperatures are more moderate than those of the northeast storms.

The greatest falls of rain occur, however, with the southeast winds, whose moisture content is greater than that of the others because they are warmer and blow off water except in Rocky Mountain districts.

South winds are warm and contain much moisture, which falls in showers rather than in continuous rains.

The southwest winds of winter precede a thaw and are much damper than west winds. In summer over much of our country they are hot, parching winds that injure vegetation.

The average velocity of the wind from these different quarters is variable in different parts of the country, the severest being on the southeast and northwest quadrants. The highest winds are always where the steepest gradients are; that is, where the barometric pressure decreases or increases the fastest. The steepest gradients are usually on the northeast and northwest sides of the storm center, with the exception of the Atlantic Coast where the southeast winds are often highest. The average for the northeast quadrant is 16 miles, for S. E. 30, for S. W. 20, and for the N. W. 30 miles an hour. But averages can deceive. As a matter of fact single instances of great wind velocities occur from each point of the compass. The greatest velocity ever recorded at Philadelphia occurred in October, 1878, when the wind blew seventy-five miles an hour from the southeast. But the record velocities for eight of the other months were registered in the northwest quadrant.

Not so high as cirro-stratus, and yet partaking of the same skeiny texture. This would be a normal sky in winter about six hours after the veil of cirrus had begun to throw its haze about the sun. No other cloud formations appear, however, and so the area of precipitation is still pretty far away. In summer such a sky is less common. If the disturbance is to amount to anything the cirro-cumulus will soon form. If the wind is from a westerly quarter the blanket of cloud is doubtless a drift from some distant storm, which will not affect this locality. The wind is always blowing toward a storm and away from clear weather.]

The period of time when the barometer is beginning to rise after having been very low is that when the strongest winds blow.

Some sections of our country have special kinds of wind that are peculiarly their own, notably Colorado, Wyoming, and Montana where the chinook reigns. This phenomenon belongs only to the cold season and only to the coldest days of it. It is a warm wind that begins to blow without much warning from the southern quarter. It is caused by a body of cold air suddenly falling from a great height. As it falls its descent heats it and it causes a rise in the temperature of the surrounding locality that greatly exceeds any rise from other causes. The increase in temperature will be as much as forty degrees in fifteen minutes.

This sudden dry heat is a great snow-eater. If it were not for the chinook the snow-blanket would stay so much longer on the cattle ranges that they would be useless as such. In northeastern sections of our country and Canada the warm winds blowing in from the ocean at the approach of a cyclone do away with the snow rapidly but with nothing like the speed of the chinook.

Another phenomenon of the air that is of tremendous benefit to man is the sea-breeze. During the intense heat of a hot wave the wind may shift to the east in Boston and in fifteen minutes coats are comfortable. Such a shift may bring relief to a strip of land two hundred miles wide along our entire eastern seaboard. The sea-breeze is explained by the fact that the land cools more quickly than the sea and also warms more easily. During the whole forenoon of a summer's day the sun has been pouring upon land and sea, but the land-air has become much hotter than the air over the sea. It rises and the sea-air rushes landward. By midnight the land has cooled off even more than the sea and the heavier air now presses out to sea again. On every normal day this balancing process takes place.

If it doesn't conditions are abnormal and chances are that mischief is brewing. This ebb and flow of warmer and cooler air is, on a small scale, exactly what is happening on a vastly larger field of operations between cyclone and anticyclone. And it is the dominance of the anticyclone with its prolonged rush of air from the northwest that interrupts the sea breeze for two or three days in winter, as the cyclone prevents the night land breeze from taking place when it is central off the eastern coast.

The exchange of air between mountain side and valley is similar to the land-and-sea breeze. The rarer air on the mountain side heats faster by day and cools faster by night than the denser air in the valley. Therefore during the day it rises and the valley air rushes up to take its place; during the night it cools and sinks into the valley. This is a great help when one is shut up in a secluded valley for several days and cannot get a good view of the skies. The atmosphere is acting properly and will remain settled so long as the air blows up your ravine for most of the day, and turns about sundown and blows out and down the ravine like a flood of refreshing water.

Of course many valleys are so large as to be affected, not by these local causes, but by the larger movements of the anticyclones when the sure-clear west wind may blow up the valley for three days at a time. But, nevertheless, for most mountainous places the logic holds and you may expect rain if the wind does not blow coolly down the ravine at night. Of course watch your clouds for confirmation.

In times of calm prepare for storm. An eminent meteorologist has frowned upon me for saying that. It is not the whole truth, I admit, but there is a certain kind of calm which happens often enough to justify the remark. It happens this way. A severe storm has passed. The customary anticyclone with its brisk northwest winds has arrived and is blowing with all the vigor necessary to induce one to believe that the clear weather is to continue for the usual length of time; that is, three or four days. But suddenly in the early afternoon, just when it should be blowing its hardest, the wind drops, lulls, shows a tendency to change its direction. There is only one explanation. Another cyclone has developed off in the west. It has knocked the anticyclone on the flank, taken the teeth out of the gale.

The wind shows this before clouds can. The absence of wind when there ought to be a lot shows it before even the first cirrus swims overhead. The chance is that when the flow of anticyclonic air has been thus rudely cut off and stillness follows, it will be storming by morning. It is best to keep an eye on these abnormal, precipitous calms. In times of peace prepare for rain.

But the eminent meteorologist was eminently right when he said that the statement was misleading unless explained. For there are many kinds of calms that do not portend coming storms. Nearly every day, winter and summer, but particularly in summer, the wind drops to a calm at sunset. That is a time of adjustment. After sunset when the accounts are all in the wind springs up with as much force as it had in the afternoon and continues until dawn. At sunrise, however, there is another truce. If this truce is neglected either at sunrise or at sunset it is a sign that either a cyclone on an anticyclone is very much in the ascendency. These truces are most often observed at the seashore when you are out sailing and the smell of supper fills your nostrils but is not sufficient to fill your sails. These calms are normal and the best sign of a fair day on the morrow, provided the other signs agree.

During the great transition period from summer to winter comes that autumnal truce, Indian Summer, which is the chief claim to fame of American weather. For day after day a brooding haze sleeps in the air, sometimes for weeks there is no wind of any strength. Winter advances insidiously in the fall but retreats in commotion, and the cooling off process permits of these still days while they are uncommon in the spring. The wind checks off more mileage in March than in any other month.

While the regular day's end calm and the calm of the year's exhaustion mean continued fair weather, there is one calm that everybody knows, which is the most dramatic moment in the whole repertory of the weather: the foreboding, ten-count wait before the knockout blow of the thunderstorm. But when that calm comes every one is already sitting tight so that it is not much account as a warning. They say that the intense stillness before the hurricane strikes is uncanny.

Whether inshore or afloat the wind is to be watched if you would know what weather is to be. It is only another of Nature's paradoxes that the most unstable element should be the most reliable guide of all on the uncertain trail of the next day's weather.

TEMPERATURES

Considering that the temperature of the sun is 14,072 degrees Fahrenheit and the temperature of space is absolute zero, 459 degrees below ours, we do very well on earth to be as comfortable as we are.

And we owe it all to the atmosphere which keeps the sun from concentrating upon us. Our place in the sun is so very small that we intercept only one-half of one billionth of the heat which it is giving off night and day. But that is sufficient to do a lot of damage if it could get at us.

But even the paltry range of temperatures so far recorded on our planet,-- from 134 degrees above zero one day in California, to 90 degrees below zero one night in Siberia,--is by no means a fair statement of the extremes we are called upon to bear. Only twice a decade in our country does the mercury vary as much as sixty degrees in twenty-four hours, and there are vast areas where the daily change amounts to only a few degrees.

The changes that do come so suddenly to us, particularly in winter and that are known as cold waves, are in reality beneficial. To them we Americans may owe our energy, our vivacity, our changeability of mood. The refrigerated, revivified air sweeping down from the north is tonic. It is heavy, and issuing from antiseptic altitudes, drives the humid, germ-nursing air from our city streets. If we had arranged a process of refreshment like this at vast expense we should have been intensely proud of it. As it is we are intensely annoyed at it and occasionally a few people are frozen to death. The Weather Bureau warnings and the coal clubs are reducing the loss in property and lives.

If you are sleeping out it is of great importance to know when the mercury is going to take one of these swoops, for sleeping cold means little real rest because one's muscles are tense, and the next day's packing needs all the relaxation one can get. Two generalizations govern pretty much every change of temperature: the mercury will rise before a storm and it will fall after one, winter and summer, but much more conspicuously in winter.

There are two reasons for this. Our cyclones usually cross our country over such a northern track that over most of the country the air drawn into them comes from the southern quarters and is therefore warmer than the air previously flowing from the anticyclone. Also the process of precipitation causes heat. This is true to such an extent on the coast of Ireland where it rains most of the time that a scientist has computed that the inhabitants get from one-third to one-half as much heat from the rainfall as they do directly from the sun. Thus a normal storm is doubly sure to warm up the environment.

In summer the reverse is partially true, for very often the rain does not begin until the actual center of depression has passed and the west winds have begun to exercise their cooling influence. So that in summer we have a sultry, sunny day as the first half of the storm area and then a cooling shower. Also after two or three days of warm weather in spring and autumn we have a rainstorm of the winter type which lowers the temperature instead of raising it. This is because the heat produced by the storm is less than that of the sun's rays intercepted by the clouds. The clear skies of the preceding anticyclone had permitted the land to warm up very fast under the midsummer sun, and the clouds of the cyclone, by cutting off the supply, had made a relative chill.

In winter the sunrays are so much feebler because of their slant and radiation proceeds so rapidly under the dry air of the anticyclone that a much greater degree of cold is produced than when the cyclonic clouds prevent the radiation. Therefore the rainy area is the warmest of all. Even in summer the winds from the southeast, south, and southwest are warmer than those from the opposite quarters, not only because they blow from a quarter naturally warmer on account of the sun, but because they are surface winds and have absorbed some of the heat from the soil. Being denser, they absorb it more readily and hold it longer.

The change, then, from the period of fair weather to that of storm brings an increase of temperature. But the rate of increase varies. The faster the storm is approaching the faster the temperature will rise; and the route of the storm's center makes all the difference as to the amount of the rise. If the wind shifts by way of the north and holds in the northeast until precipitation begins the rise in temperature will be very slight. The great snowstorms of the northern half of the country occur under just such a circumstance. If the wind shifts by way of the north but gets around to the east or even southeast before the precipitation starts the rise in temperature will be more pronounced, as much as thirty degrees sometimes in a few hours, and the winter storm that started in as snow soon changes to sleet and rain.

If the wind shifts by way of the south and then into the southeast the rise will be vigorous and the storm will likely be a comparatively warm rain. If the wind shifts only so far as the south the rise will be highest of all and blue sky will often appear between the showers, showing that the air is heated to a considerable height.

The progress of the temperature changes from the maximum of the cyclonic area to the minimum of the anticyclone is also dependent upon the wind. If the storm center is passing south and the wind begins to pull into the northeast and north the temperature will fall steadily and slowly. The rain or snow often cease gradually by the time the wind has reached the north, but the temperature continues to fall slowly until it reaches very low levels in mid-winter. If the storm center is passing north of you the wind which has brought most of the rain while it was in the southeast with comparatively high temperatures swings into the southwest, the temperature falls somewhat.

There is usually a final downpour and a rapid shift of the wind into the west or northwest, but almost never directly into the north. The temperature falls several degrees in a few minutes, quite unlike the gradual decline of the northeast-by-north shift, and clear skies come at once with rapidly diminishing temperatures. In the vicinity of Philadelphia a fall of twenty-five degrees would be most unusual on the northeast shift,--such storms reaching 38 degrees and falling to 15, while with the other shift a fall from 55 degrees to 15 would not be unusual. Of course any one set of figures given could only show the tendency and not the rule or limits.

After the manner of the wind-shift the intensity of the storm is a good gauge of the temperature change to be expected by the camper. As a rule the greater the intensity of the storm the greater will be the degree of cold that follows it. The storms that have a complete wind circulation about them are always more severe than those with incomplete circulation and are invariably followed up by some reduction in temperature. If the decrease is not proportionately great and the subsequent wind has only a moderate clearing quality look out for another cyclone.

In such a case the temperature is the best witness of the contemplated change. For instance, after a summer thunderstorm a decided coolness is de rigeur. If this does not occur it means nearly every time that there is another thunderstorm in process of construction. There may be not a cloud in the sky, there may be no wind (although there should be) so that the course of the thermometer is the only means of telling what is to be the next event. Anybody can take a thermometer with him although a barometer--the most accurate forecaster of all--may be thought too much expense and bother.

At some future date the Weather Bureau will be able to predict the temperature of seasons in advance. This, together with the amount of rain scheduled to fall, will be an invaluable aid to everybody and to the farmers most of all. At present mild seasons that have severe storms without the appropriate degree of cold after them cannot be entirely explained, let alone being prediscovered. They all hinge upon the more or less permanent areas of high and low air pressure over the oceans and international meteorological service has not progressed far enough to support many ocean stations as yet.

Sometimes clear weather may intensify, growing brighter, stiller, colder. This is because the pressure is increasing. Cold seasons are distinguished usually by a succession of anticyclones. There is no way of telling how long a certain spell of cold weather is to last, but I have noticed that the same characteristics rarely predominate for longer than a month at a time. In other words, if December has been warm and rainy, January will likely be cold and dry. Of course, that is precisely the unscientific sort of generalization which the Bureau very rightly frowns upon, but which one may nurse privately until science has provided a substitute as she already has in so many instances.

With a little practice it is an easy matter to estimate the temperature to within a very few degrees. Try guessing for a few mornings and then look at the thermometer. You will hit within three degrees every time after a week of this.

Allowance must be made for the amount of moisture in the air and for the force of the wind. Damp air feels colder by several degrees than crisp, dry air, and a breeze increases the difference still more. Air in motion is not necessarily colder than calm air. As a matter of fact the lowest temperatures of all are recorded about sunrise after a still, clear night. The amount of radiation accomplished during the last hours of the night is amazing, and the downward impetus of the thermometer is often carried on for an hour or more after the sun has appeared above the horizon. A self-recording thermometer is an amusing toy which will show this and becomes a valuable instrument if one raises fruit.

In winter three o'clock of an afternoon sees the highest temperature usually, and in summer this maximum occurs as late as half-past five, due to the fact that the sun can pour in its heat faster than the earth can radiate it off. For the half hour before and after sunset, particularly in winter, the loss of heat is relatively greatest; then the pace slackens till three or four in the morning, when the plunge of the mercury is accelerated until the rays of the rising sun counteract the radiation.

If the mercury does not rise appreciably on a clear winter's day it is a sign that a cold wave is stealing in, due, doubtless, to a gradual increase in pressure without its customary bluster. Very often snow flurries predict its approach, but this may be so gradual that only the restriction of the daily thermal rise may indicate it. By the next morning the temperature will likely be twenty degrees colder.

If the mercury does not fall on a clear winter's night it is a sign that a layer of moist air not far above the surface of the earth is checking the normal night radiation. Unsettled weather is almost sure to follow unless this wet blanket is itself dissipated and the mercury takes its customary tumble before morning.

If the temperature falls while the sky is still covered with clouds clearing, possibly after a little precipitation, will soon follow.

Hot waves approach insidiously. A night will not cool off as it properly should, the sun will rise coppery, and while the day is yet young everybody begins to realize that all is not exactly right. But the heat increases usually for several days, not only by reason of steadily lowering pressure, but also by accumulation. Finally when a climax is reached it departs abruptly on the toe of a thunderstorm.

A cold wave reverses the process. It arrives abruptly on the heels of a departing cyclone and, after losing power, steals away without any commotion whatever. Its rate of progress is in close relation to the cyclone ahead of it.

Our mountains play a great part in our weather. They are a right arm of Providence to our agricultural communities. Due to their north and south trend a cold wave of any severity reaches the Pacific Coast only once a generation. Just once has snow been observed to fall at San Diego and it is so rare south of San Francisco that many people never have seen a flake. East of the mountains the belt of desert makes natural crops impossible for a thousand miles, but if they crossed the continent all the territory north of them would have such a cold climate that none of the present enormous crops of Canada and our northern states could possibly be grown. It is also due to the wide insweep of winds from the Gulf that the plains states are so well watered.

The tops of cumulus are irregular, looking like wool-packs; the bases are flat. The true cumulus shows a sharp outline all the way round. Its shape is in constant change due to the strong winds it is encountering. It is caused by the swift uprush of warm air on a sunny day. This cloud is a sign of fair weather, because the base is not large, compact, or dark enough to threaten rain and its comrades are also disjointed. If the cumulus grow darker toward the horizon and increase toward evening a squall is likely.]

In lesser fashion the Appalachians protect the Atlantic seaboard. They withstand the impact of the cold waves to a great extent, although they are not high enough to divert the flow of cold air entirely toward the south and it is not desirable that they should. As things are the cold strikes Alabama before it hits New Jersey, and is often more severe there.

Comparative cold is often registered by the green color of the sky. A fiery red continues the prevailing heat.

The day that is ushered in by a fog, in summer, will likely be warm, providing the fog lifts by ten o'clock.

The temperature of a night with even a thin covering of clouds will be a good deal higher than if the sky is clear. In the British Isles the whole difference between freezing and no freezing lies with the fairness of the heavens. Everywhere frost will not form while the sky is covered, although the temperature may be below the freezing point. In summer radiation on a still clear night may be so rapid that frost may follow a temperature of fifty degrees at nightfall.

The temperature at the surface of the earth may easily deceive, as a colder or warmer stratum of air may overlie that immediately next to the ground. I have seen water particles fall when the temperature was as low as 16 degrees above zero, showing that the stratum of cold air was very thin. Our sleet storms in which immense damage is done to trees and telegraph wires occurs from just such a situation,--a cold, shallow layer of air close to the earth, with the warm moisture-bearing air flowing over it. The reverse of this situation is not uncommon--the sight of a snowstorm proceeding merrily along with the ground temperature at 35 or even 40 degrees.

Coming warmth may be noticed by the increase in size of snow flakes, with finally hail and rain. Coming cold is foreshadowed by hail mixed with the rain and lastly snow flakes which have a tendency to decrease in size. Colors of the clouds predict temperature changes, but it takes much practice to distinguish the cold, hard grays from the soft, warm ones. A warm sky is always less uniform in color than a cold one. The colors of winter sunsets are, as a rule, much brighter than those of summer skies.

The stars seem brighter on a night that is to be cold. If they twinkle it is because of rushing air currents, and if the wind is from the northwest the result may be a subsequent lowering of temperatures.

The whole question of whether it will be colder and how much is vital to the camper and if the signs of change are taken along with the look of the clouds and the direction of the wind he need never be wrong as to the direction the mercury is going, and will soon be able to guess the distance pretty fairly.

RAIN AND SNOW

East of the Mississippi River rain falls with the utmost impartiality upon every locality. Thirty to fifty inches are delivered at intervals of three or four days throughout the year. And if there is a slight irregularity in delivery one can be sure that from 125 to 150 of the 365 days will be rainy. Occasionally there is a more or less serious hold up of supplies, but this rarely happens in the spring of the year and never happens to all sections at once. And if there is a desire to make amends for the drought, we have what we call a flood and blame it on the weather instead of on our precipitous denudation of the watersheds.

West of the Mississippi particular people have to go to particular places for their rain. If they like a lot of it they must go to the coast districts of Washington or Oregon where they can have it almost every day. It rains a good deal at Eastport, Maine,--about 45 inches a year; that is, nearly an inch a week,--but at Neal Bay, Washington, at about the same latitude, in one year it rained 140 inches, and it never stops short of 100 inches any year.

On the other hand, if the Washington people are tired of it they need only escape to Arizona where it rains about two inches a year, and they can live in an enterprising hotel down there whose manager believes that it pays to advertise the sun. He guarantees to provide free board on every day that the sun doesn't shine.

In the plateau section enough snow falls every year to store up enough water for irrigation purposes, and the little rain that falls arrives in just the right season to do the most good, the spring. In California what the farmers lose in amount they make up in the regularity of its arrival.

North of the Ohio River most of the precipitation from November to April is snow. About 50 inches of it falls on the average over this tremendous territory. And it is more useful than rain,--the handy blanket that makes lumber-hauling easy, that keeps the ground from freezing to Arctic depths, that fertilizes the soil, and that acts as a great reservoir, holding over the meat and drink of the vegetable kingdom till the thirsty time arrives. In upper Michigan and Maine the average depth becomes 100 inches. Averages are very misleading when snowfall is being considered, some winters producing very scanty amounts and others heaping it on to the depth of 185 inches once at North Volney, New York.

South of the Ohio the depth varies from substantial amounts in some winters to almost nothing in others. Snow has been observed, however, in every part of our country except the extreme southern tip of Florida. Once and only once on the records a great three-day snowstorm visited all of southern California, extending to the Mexican border and to the coast.

The strip of country between the parallels of New York City and Richmond comprises the section wherein each winter storm is one large guess as to whether the precipitation is to be snow or rain. A compromise is usually affected in this way. Before the clouding up began the mercury may have stood at ten degrees below zero. As soon as the wind acquired an easterly slant the temperature increased. As it neared the freezing point the snow would begin, first in flakes of medium size which would enlarge until after a particularly heavy fall of a few minutes they would at once almost cease. Hail soon would succeed, the mercury still rising, and often the hail would have turned to rain before the freezing point of the air of the immediate surface of the earth had been reached, turning the snow already on the ground to slush and making a holiday for germs.

One can always tell when this change to warmer is about to occur because the clouds which have been part and parcel with the obscuring snow suddenly show, not lighter but darker. The sudden increase in size of the flakes is another infallible symptom of increasing warmth in the atmosphere for each large flake is a compound of many smaller ones. When the temperature is low the flakes are very small, being grains and spicules in the severe blizzards of the west and falling as snow-dust in the Arctic. In the heavy storms of the guessing-belt the flakes are not necessarily small.

I have noticed (in the latitude of Philadelphia) that our largest storms begin very leisurely indeed with small and regular-sized flakes. A quarter of an inch may not fall in the first hour. As the center nears the snow comes ever faster and larger, but not large, flakes are mixed with the original-sized flakes. Snow dust is apparent. At the height of the storm flakes of all sizes except the very large are falling, denoting great activity in the strata of air within the storm influence. In the ordinary storm an accumulation at the rate of an inch an hour denotes a storm of considerable intensity.

The snow will likely keep on falling as long as the flakes are irregular in size. If they grow large and few or very small a cessation is likely, even though the wind is still blowing from an easterly quarter. The amount of snow likely to fall can be gauged not only by the process of flake-change but by the rate at which the wind rises. A storm's intensity is measured by the amount of wind. A storm can be a storm without a drop of rain or flake of snow if only there be enough wind. And as long as the wind in a snowstorm keeps rising the storm is likely to go on, probably increasing in volume of precipitation.

If the wind shows a tendency to edge around to the southeast there is danger of the snow turning to rain; if the wind veers slowly to the northeast the temperature will fall slowly and the rate of precipitation will likely increase for a while. In such instances the snow does not continue to fall after the wind has swung west of north. Often clearing takes place with the wind still in the north or even a point east of north.

Contrary to superstition snow may begin to fall at any hour of the day or night. But certain hours seem more propitious than others, owing no doubt to the tendency of cooling air to condense. Three o'clock of an afternoon and eight o'clock in the morning are favorite times, the one being the hour of a winter afternoon when cooling is begun, the other the hour when the coldest time is reached and condensation likely if at all. Of course, one remembers storms beginning at nine, ten, eleven, and every other hour.

Storms that begin in the morning seldom reach much activity before three o'clock in the afternoon, while those that begin then quickly increase in intensity as evening draws near and the sun's warmth is withdrawn from the upper air-strata. More snow falls at night than in the daytime, also. Snow is more delicate than rain and perhaps more responsive than rain to the subtle changes of the atmosphere. Possibly there is no ground on the Bureau records for these ideas, possibly storms have a tendency to start from the Gulf on their northeastward journey and so reach Philadelphia oftener at one time than another. I would like my notions confirmed that snowstorms increase at nightfall, and that they prefer to start operations at sunrise and about sunset.

For the camper the snowstorm need have no terrors. It gives a long warning of its approach. It comes mostly without destructive winds. Its upholstery protects and warms the walls of one's tent. It adds beauty to the leafless woods, interest to the trailer, and a hundred amusements among the hills.

But the value of snowy weather is not only measured by its beauties and commercial uses. There is another way: make it read character for you. Watch the reactions toward the first snowfall of half a dozen kinds of people. It will show you what they are; give you a very fair measure of their youth.

Our atmosphere contains a lot of moisture that never gets precipitated. You can prove this on any warm day by noticing the way the atmosphere acts toward a glass of ice-water. When the air of the room is much warmer than the surface of the glass it surrenders its moisture willy nilly. Sometimes this condensation is enough to cause a miniature rainstorm that trickles down the outside of the tumbler. If a small cold surface can wring so much water out of a little air it is small wonder that we get an inch or so of rain from vast currents of air at unequal temperatures.

Try to visualize the process. A stream of vapor has been warmed and is ascending. A mile up and it has cooled not only by the reason of altitude but also by the process itself. About each little dust-particle in the surrounding area vapor forms--vapor cannot form without something to form on, there being always enough dust from deserts and volcanoes to go round. If the cooling proceeds the tiny globules enlarge and as they increase in weight they settle and fall. Falling, they unite with others.

If the air-strata are very warm and thick the drops may grow to a very considerable size. We see these in the middle of our great winter rains when the insweep of southern winds with all their warmth and moisture is very extensive. Also the first few drops that come from the thick, hot lips of the thundercloud are usually immense.

The best way to measure the size of a raindrop is to have it fall in a box of dry sand. It rolls up the sand and measurements can be easily and accurately made. But the most interesting way is to let the first drops of the thunderstorm fall upon a sheet of blotting paper. If the same sort of blotting paper is used the measurements will be of just as much importance for comparison. Circles as big as teacups are formed sometimes.

Heavy drops in winter mean a heavy fall, because they denote high temperatures which are uncommon and are bound to be followed by considerable condensation as the cooling proceeds back to normal temperatures. Small drops in summer mean either cooler weather, or sudden condensation. Small drops in winter are a sign of very thin moisture-bearing strata, or low temperatures, indicating that the rain will be light, protracted, and liable to change to snow.

Hail is frozen rain. Winter hail is small and harmless and rarely falls to any depth because the exact temperatures that bring forth the hail rarely continue for very long at a time. Hail in winter is merely the stepping stone to either rain or snow. But in summer hail is a serious matter. It shows that there is a violent disturbance of the atmosphere in progress. Vertical air currents, probably abetted by electricity,--the authorities are not sure--often carry the stones up several times. They take on layer after layer, coalesce, and sometimes fall the size of eggs, apples, or any other fruit, barring melons. The usual summer hail does not exceed the size of a robin's egg. Even a projectile of that size, however, falling for a half mile or more has a tremendous destructive power. Greenhouses suffer, birds are killed, cattle stunned, and loss of life has been known to follow. In August in 1851 in New Hampshire hailstones fell to the weight of 18 ounces, diameter 4 inches, circumference 12 inches. In Pittsburgh stones weighing a full pound have crashed down, and in Europe where many destructive storms have occurred there are official records of even greater phenomena. The lightning accompanying these hailstones is usually very severe. A flake or ball of snow forms the nucleus of a hailstone.

If a thundercloud looks particularly black or if it can be seen in commotion think of hail and seek shelter. It is pretty difficult to predict exactly when hail is going to fall in summer. It is a possibility with every large storm, but a probability with only a very few during the summer. It accompanies tornadoes.

In winter hail falls before a rainstorm, even when the ground temperature precludes the possibility of snow; some lingering stratum of cold air has ensnared the drops on their way down.

Snow is not frozen rain. It has an origin of its own. It is born in a temperature consistently below freezing and on the condensation of the invisible moisture becomes visible as a tiny crystal. These infinitesimal crystals unite and form larger, hexagonal shapes, elongated or starry. They are wafted along, sinking, all slightly differing one from another, although forming a few types. These types have been photographed and catalogued and very often the altitude from which the snow is coming may be learned from their shape and design. But this branch of science is young yet and confusing and the outdoor man has surer signs of the vicissitudes of the storm, in the general size of the flakes, the power and direction of the wind, the clouds and temperature. The possibilities of flake-study as a means of forecasting are many and of value as is anything that tends to unveil the secrets of the greater heights.

Snowflakes are so light that after the storm processes are over and the sun has come out the residue may still float lazily to the ground.

The wild disorder of the snow flurry will only last a few minutes and never leave much snow on the ground.

Snowstorms that come on the wings of the west wind may be severe, but they will be short. They are unusual in the east, but sometimes the heaviest snows of the western states come on the sudden cooling that follows the shift to west.

Snowstorms arriving on a high wind last only a few hours.

Snowstorms that are long in gathering and increase to considerable intensity continue a long while.

Those that follow a sudden clouding up are of no importance.

The snowstorms that leave on a high wind from the west or northwest are followed by a cold wave. Those that continue after the storm wind has died away are succeeded by calm, clear, and usually warmer weather.

In northern districts a snowstorm may be looked for after a period of cold weather. In middle districts if the cold has been severe the reaction to warmer may bring rain instead. In such cases generalities are of no use, and the possibilities must be determined by the man on the spot. The best conditions for snow through the middle districts are occasioned by an area of low-pressure with its attendant precipitation crossing the southern half of the country while the northern half is under the influence of an area of high-pressure with its attendant frigidity. The cold air flows into the southern storm with the result that the middle districts get the northern quadrants of the storm which are the usual snow-bearing ones instead of the southern rain-bearing quadrants that they would have got if the center of the storm had pursued its usual course up the Ohio and down the St. Lawrence.

If the storm has two centers, one over Texas and the other over Montana, as is so frequently the case in winter, the subsequent high pressure will come too late to affect the temperature of the zone of precipitation and the latter will likely be rain in the middle districts. Sometimes the cyclones cross the country on the Canadian border and enough warm air is sucked over the line to give the inhabitants of Montreal a thaw and rain. This happens to them only once or twice a winter. And even more rarely a cyclone over the Gulf with an anticyclone above it will give the Gulf States a taste of winter, but rarely more than a few flakes.

It really all depends on the influx of air, its rate and direction. It rains in

Alaska and snows in Georgia on the same day merely because at one place the air is coming off the Pacific, and at the other it is flowing from the center of a refrigerated continent.

And the progress of these storms is one of Nature's greatest poems if you take a minute to think of them sweeping on in majesty, the one thing that man cannot control. Even the snow which is the citizens' curse as well as the farmers' blessing becomes epic when it beleaguers an empire for half a year.

DEW AND FROST

The very process that made the tumbler of ice-water sweat on the hot day causes dew. And the formation of frost is analogous to that of snow. Frost is not frozen dew, but the formation of moisture crystals at the temperature of 32?or below. Frost or dew form only on still, cloudless nights. Even if no clouds are visible, neither will form if a stratum of humid air has prevented radiation. Hence either dew or frost is a fairly good sign of clear weather.

Three white frosts on successive mornings are followed by a rain. This saying holds water not because there is any virtue in frost to cause rain, but because a storm is normally due once a week. The frosts did not form when the anticyclonic winds were blowing and usually not more than three mornings elapse between the time that the anticyclone has lost its influence and the time for the next cyclone to appear. Frost indicates a considerable amount of moisture in the atmosphere, also, which tends to increase as the cyclone approaches.

The heaviest dews come in late summer and the heaviest frosts in mid-autumn because the change in temperature is greatest then and there is a greater chance that there will be a calm at sunrise. The greatest frost damage occurs in the spring because the tenderer crops are growing then. Summer frosts used to occur in the northern parts of Minnesota and along the southern boundaries of the inland Canadian provinces before the forests were cleared off. The march of civilization has actually pushed back the frost line some distance.

Frost may occur when the amount of humidity in the air is low and the barometer rising at any temperature under 50 degrees at nightfall, the clear skies permitting radiation enough under those circumstances to produce the necessary cooling. An evening temperature of 40 degrees with the clear skies and faint west breeze will almost surely produce a frost, provided the wind drops. In such circumstances the only hope for the farmer is that there is enough humidity in the air to cause a fog before the frost-point is reached. A temperature touching 34 degrees would not bring frost, however, if the sky was at all overcast. Frost is difficult to predict because a night shift in the wind, cloudiness that forms after midnight, or even a wind arising before the coolest period at dawn will prevent its formation. On the other hand, clouds may disperse, the wind may fall or radiation may be so rapid before sunrise as to cause a killing frost unawares. The farmer who lives in areas disputed by winter and spring may never be quite sure, but precautions should be taken on the still, clear, dry nights with the thermometer at fifty or below.

Fruit-growers resort to fires or to coverings to protect their crops. The fires are particularly worth while, not so much for their heat which at best cannot be expected to warm up the great outdoors much, but for the smoke which prevents radiation. A line of smudges such as campers use to ward off the mosquito would spread a pall of smoke over an orchard efficaciously. A snowstorm, the soft fluffy sort that falls in April or May, can do much less damage to vegetation than a severe frost.

Temperatures are much lower on the ground than even six feet above the grass. Naturally these temperatures are those that really influence most vegetation and in England temperatures on the grass are given in the weather report with the ordinary observations, being as much as six or eight degrees lower on clear nights.

In some of the hot, dry countries, such as Arabia and Egypt, most of the moisture that they receive falls in the form of dew. Falls, of course, is a loose expression as the dew forms and does not fall, being different from the minute particles of fog. The fog particles in suspension in the air are estimated to be as small as 1-180th of an inch. When they grow to 1-80th of an inch in diameter they commence to fall. Fogs are chiefly caused by the soil being warmer than the air above it; the vapor on rising condenses and becomes visible. In the spring and fall currents of air blow over rivers at different temperatures and the result is a fog. One does not have a fog in the desert.

There are places in the ocean with cold and warm currents with the air above them correspondingly different where fog is of almost constant occurrence. The Gulf Stream off the Grand Banks of Newfoundland has a temperature of 78 degrees, while the water on the Banks is 45 degrees so that fogless days are rare along the line of meeting.

Frost is known in every part of our country, many localities in the plateau section being exposed to it every month of the year. The thin air and cloudless skies of the altitudes make radiation very easy and the daily variation of temperature is much wider than along the humid coasts. Those who have never looked into frost conditions throughout our country will be surprised to read the warnings of the Weather Bureau.

From the station at Pensacola, Florida (frost-proof Florida!), comes this statement: "Vegetables are subject to damage by frost during all seasons of the year."

Pittsburgh, Pennsylvania, "Frost is likely to damage fruit or other crops in May and September."

Phoenix, Arizona, "Frost is likely to do damage in December, February, and March."

Baker City, Oregon, "Fruit and other crops are most liable to damage by frost in April, May, June, September, and October."

Kalispell, Montana, "Frost damage for fruit, May 15th to July 10th; for grain, June 25th to August 1st."

Montgomery, Alabama, "During March, April, and May fruit and early vegetables are subject to damage by frost."

THE THUNDERSTORM EXPOSED

Probably nothing in the world causes more terror than a flash of lightning. In an able-bodied thunderstorm playing about a city there are several dozen flashes, and every one of them brings trepidation, fright, or positive terror to thousands of human beings,--oftenest women, sometimes men, and occasionally children. Yet probably there is no alarm in the world so ill-founded.

Thunderstorms play pretty generally over our three million square miles with their hundred million population. Yet lightning picks out of this crowd only three hundred people a year who are foolish enough to be killed. That is, only three persons in each million to be sacrificed to the most astounding and beautiful display in the world, a mere handful compared to the mounds of motor car victims or to the 33,068 deaths a year attributable to railroads and the perils of track-walking.

The trouble about the thunderstorm is that it does not lull one into the sense of insecure repose. It is too obviously after one. If the thunder were toned down a bit and the lightning a trifle duller the alliance might claim its thousands, like the inconspicuous housefly, and never meet an objection. But until the thunderstorm foregoes its bravado it will continue to bully the ladies into hysterics.

Of course, there is always the sporting chance that you are one of the three in your particular million to perish.

But you can lessen the chance. You must not seek refuge under a tree. You should not take doubtful shelter in a barn. And you had best not sit in a draft by an open window if there is a tree just outside it. By these three avenues most of the thoughtless three hundred (a year) invite their end.

Trees that are tall and otherwise exposed are struck oftenest. The electricity in the cloud and the electricity in the earth are always endeavoring to combine. When this tendency becomes so strong that the resistance of the intervening air is counteracted the electric discharge between thundercloud and earth takes place. This happens most frequently from some pointed thing as a steeple, a tree if they are good conductors. Men and animals are sometimes charged with the electricity opposite to that of the cloud. When the lightning is discharged, even at a distance, the bodies revert rapidly from the electric to the natural state. This return shock or concussion occasionally proves fatal.

That is the reason that trees are such poor protectors from the storm's fury. Better a wet skin in the middle of a field than precarious dryness under an oak or cherry or tall pine or almost any other tree. If it should hail hard enough to stove in your head take to a beech or a small spruce.

Barns are struck so often because the body of warm, dry air in them favors the passage of electricity. Those who hide in barns are sometimes cremated. After a severe thunderstorm in the Poconos I have seen as many as three barns on fire at once.

Open windows, porches, and exposure generally are safe, but not safest. The cellar, that old stamping ground, is where instinct takes a few. Any closed room on the side of a house away from trees is good enough. But the risk of annihilation is so very small that one is repaid for taking it by the spectacle. A great thunderstorm surpasses anything in nature in the matter of architecture, coloring, directness, and surprise,--which, with selection, comprise the essentials of art. Imagine the crowds that would pay to wonder at the sight if a thunderstorm could be staged, say, at the Hippodrome!

Some hot morning, if you have time to watch, you may see a thunderstorm born in the mountains. The warm, moist air flows up the mountainside and the essential start is made. Cooling, this air first shows as a fluffy cloud that soon grows harder in appearance and becomes tufted at the top. Its little belly swells and grows blacker. It hovers over the valley. Others add to it. Suddenly a sort of adolescent thunder is heard. The tension has become too great. A definite consolidation is visible, a fringe lowers, and a few drops of rain may reach you.

The incipient storm moves off, and having started a whirl within itself, increases, like a rumor, as it goes. Before it has moved beyond your horizon it may have become a large patch of dark blue with billowy white crests on the top, and underneath hangs a curtain of rain. Chances are that it will not go far before encountering conditions that dispel it, but it may cover half a dozen counties before nightfall. As a rule these little heat thunderstorms do not amount to a great deal. They are originated by local conditions and leave things pretty much as they found them.

But when a cyclone is passing in summer a series of thunderstorms or heavy showers with some thunder frequently take place instead of the all day winter rain. These thunderstorms mount up against the wind. Their clouds are black. The word black is an indulgence of the human weatherman meaning, of course, any dark color,--a black sky would terrify the most hardened of meteorologists.

The cyclone winds come from the south or southeast just as they do in winter, but this quarter may not bring the heaviest rainfall in summer. There may be showers or even clear skies, but the day will be humid and hot. A haze of cirro-stratus cloud will gradually overspread the sky from the west, darkening into a blue from the original whitish or gray. Lightning does not appear from the cirrus, but after the sky has grown pretty dark a ridge or tumbled cloud will be seen low on the western horizon. Meanwhile the wind will have died down.

The lightning, at first only a faint glimmer, will have become more frequent and noticeable. If it is striking at a distance of fifteen miles the thunder will not be heard. As soon as the storm center, where the heaviest rain and the electrical display are taking place, gets within the fifteen-mile radius thunder will be heard to growl, and the tumbled cumulus clouds which may have lain along the horizon for hours will begin to approach. The storm will be upon you in ten minutes likely after the arc of foreboding blue and white cottony cloud has begun its charge across the sky. Light quickly fades from the heavens. The wind drops entirely. Streaks of lightning burn downward.

Behind the arc stretches a curtain of uniform blue or gray. If the gray is lighter in places the rainfall will not be heavy. If the curtain is a uniform blue a heavy rain is sure. If the bow of clouds can be seen to tumble or is continuous and approaches fast the wind is certain to be severe,--may be from 30 to 60 miles an hour for the first few minutes. Sometimes a cloud of dust advancing before it demonstrates its force.

This moment immediately before the storm breaks is the dramatic moment of the entire cyclone. As in a tragedy, the interest has built up to this supreme occasion, this knife thrust, from which interest recedes until clear skies show that the play is over. From 12 to 36 hours is the usual time required in winter. In summer the cyclone takes even longer to pass a given point, but the period of rainfall, in which the winter storm's amount is often surpassed, may not last fifteen minutes. First the blow, then a crash of thunder, and the rain in big drops, which lessen rapidly in size as the whole world seems involved in the vast forces of the storm center. Most of the precipitation occurs in the first fifteen minutes, sometimes in the first five. A hearty storm will deliver an inch in short order. Although the rain continues often for an hour and sometimes in the storms that are attached to a well-defined cyclonic system there will be two or three robust thunderstorms in succession, yet the first downpour is usually the torrential one and the others die away until the conditions that caused the outbreak have passed off. With the severer storms hail falls.

The general condition of the air after a thunderstorm is cooler, dryer, and more invigorating than before. Ozone has been liberated, dust has been washed from the air and vegetation. The surest sign of a continuation of unsettled weather is the failure of the atmosphere to cool off. If the air remains sultry and heavy and depressing another shower is due. In such circumstances the wind will not have begun to blow with any great promise from the west.

A close, sultry morning is the best indication of a thunder-gust. The large piles of cumulus clouds are called thunderheads for the very reason that they almost always precede a thunderstorm. The heaviest electrical disturbances have cirrus clouds a few hours in advance of them very much as their winter relatives. A thunderstorm that does not cause the barometer to fall considerably will not amount to a great deal.

At night the different kinds of lightning furnish a running commentary to the storm. On calm evenings the sky will be cloudless, with perhaps the exception of a low rim on the northern horizon. Yet flashes of lightning, of course without thunder, may be seen illuminating that entire quadrant of the sky. This is called heat lightning and is popularly supposed to be the result of the heat only. As a matter of fact it is caused by a normal thunderstorm that is operating below the horizon. Reflections from this storm are shown on the rim of clouds, or if no clouds are visible, on the bowl of the sky. If you see lightning be sure that there is a storm somewhere.

If this disembodied sort of lightning continues to flash from the western sky it is quite possible that the storm will reach you. If it shows on the northwest or north of you the chances are that the storm will be carried around. If the wind is from the southwest and the lightning appears there only the progress of the clouds will show whether the storm is pursuing the normal track from the west and around you or whether it is edging up toward you. One cannot be very well surprised by a thunderstorm of any energy in camp as the lightning shows as much as two hours before the storm breaks and the thunder gives fifteen minutes' notice on most occasions.

The sort of lightning that spends itself illuminating the clouds in serpents and willowy branches confines itself to the altitudes and is very beautiful and harmless. It is accompanied by thunder that sounds hollow, that rumbles over the sky, and usually does not end with the crash and thud of the more vigorous variety. Such lightning and such thunder are more often connected with the sort of storm that comes up very swiftly on a western wind. It gives shorter warning than any other sort of thunderstorm and is not connected with the cyclonic area. I have known such a storm to manifest itself low in the west, approach, and break within twenty minutes. Much wind results and not much rain, although the temperature falls. Lightning with storms of this impromptu kind rarely does any damage.

But if the storm rises slowly against the wind, requiring an hour or two or three to approach and break, the lightning will grow almost continuously, some of the flashes being broad streamers cleaving the western sky. It is this sort of lightning that does the damage. The thunder, instead of rolling like an empty barrel, hits into a series of concussions. If the lightning strikes an object nearby the crash is rather appalling. There are several freak sorts of lightning such as the ball form, which are rare.

The approach of the center of disturbance may be gauged by the length of time that elapses between flash and crash. In reality the thunder occurs immediately after the discharge of electricity, but sound travels so slowly, compared to light, that a minute may intervene between stroke and clap. You may count the seconds, noticing the regular decrease, signifying the nearing of the crisis. Soon a flash in front and a simultaneous peal will show you that you are in the thick of things. The next bolt or two may hit very close and you can appreciate what it means to be on the firing line. Then the next river of fire with its detonation streams behind you and you are saved.

In a severe thunderstorm there are several centers, several nuclei that shed destruction like great batteries and their progress over and beyond you has its thrills. You may find the exact number of feet away that the bolt hit by multiplying the number of seconds elapsing between the lightning and thunder by 1120. But an easier way is to allow a mile for every five seconds on the watch. One or two seconds, and you are pretty near the center of the fray.

Lightning compresses the air, leaving a partial vacuum. The other air rushing in to fill this partial vacuum forms the wave motion that produces the noise. That is the whole why of thunder. The reason thunder rolls is that the lightning is a series of discharges each of which gives rise to a particular detonation. If lightning were but one discharge, the thunder would be but one stupefying crash. Reflections from the clouds and from layers of air of different densities and from the ground are agencies that prolong the sound.

Our atmosphere is never lacking in electricity. This electricity is always positive in clear weather and sometimes negative in cloudy. Science concludes, then, that negative electricity invariably indicates rain, hail, or snow within a radius of forty miles.

Moist air is a good conductor. Our powerful motors can now produce a spark of electricity several feet long. But some of the flashes that shoot across the sky in a big storm extend over five miles. The duration of the flash varies from 1-300th of a second to a second. The reason that lightning does not always pass imperially along a straight line is that some air, either moister or warmer than the air around it, offers less resistance. The lightning takes this line of least resistance along the pathway of warmer or less dense air.

Altitudes of thunderclouds vary. They may hover above the earth at 800 feet. They may be a mile high. They have been observed on peaks of mountains three miles high. Many other electrical phenomena are observed in the mountains. The study of these will undoubtedly benefit meteorology and perhaps go far to explain the unsolved problems of the Service.

One kind of thunderstorm that is rather rare is that which arrives in winter with the passage of an energetic cyclone. Often when the wind, having been in the southeast for most of the storm, is passing around and reaches the south or southwest the rainfall culminates in a deluge and thunder is heard. One or two such storms are a winter's complement. They usually terminate the rainfall for that particular cyclone. I have never heard of damage caused by these winter electrical storms, and they occur only in exceptionally well-developed areas of low pressure.

Lightning has many times been observed during heavy snow storms. I have never heard any thunder with it. The discharge must have been very faint.

Stratus is merely lifted fog in a horizontal form, the lowest of all, and the simplest as regards structure. It means neither rain nor snow and the apparent clearness of the blue above it would indicate clear weather to come. But through the break in the stratus near the horizon shows a cloud of firmer texture, which is less reassuring. Stratus over the land in winter takes the appearance of long bolsters of gray through which a pale blue sky shines. Such clouds may blanket the sky for days without causing a drop of rain. If they show a tendency to glaze over expect snow or rain, but not in large quantities.]

The fascination that a thunderstorm has for many people is explained partially by the fact that one sees the whole process from beginning to end. The officials of the Weather Bureau have this privilege as regards cyclones. It is their business and pleasure to watch the setting up of these vast storms, to follow them on their journey. It is small wonder then that they find the spectacle fascinating.

THE TORNADO

The birds, the flowers, and the tornadoes are all busiest in spring. And the tornadoes probably make the largest impression.

A tornado is merely a whirl of air, caused, as are all the other whirls, by a striking difference in temperature in adjacent areas. A tornado is a local and restricted example of the same thing that a cyclone is. But a tornado rarely crosses more than a single state; a cyclone strides continents. A tornado lasts, in one place, about a minute; a cyclone affects the weather for three days. A tornado never survives the night; a cyclone plods on for a week. And yet if you are betting on destruction put your money on the tornado. What it lacks in the realms of space and time it makes up in intensity. Its sting is fatal.

Tornadoes occur chiefly in the spring because the temperature changes are greatest then and it is from these that the tornado sucks its nourishment. Over the plains, for example, a limited area is abnormally heated by a local cause. Abnormal cold comes in contact with the abnormal heat. The great difference in pressure results in a spiral as it did in the cyclone, only in a very small spiral, and once begun its energy is self-aggravating. The whole thing moves off toward the northeast attended by the black cloud of its condensation. From the black cloud a funnel like an elephant's trunk sways back and forth, now touching the ground and now escaping it. The black cloud has been in the southwest for some time probably before it has commenced to move. The day has been very oppressive. The sun rose rather coppery, in all likelihood. As the black cloud with the swaying funnel nears a roaring is heard. Darkness falls. The roar increases.... Instantly it is over.

Now that you've been through a tornado you know how it feels,--almost. After the funnel passes hail falls, lightning flashes through the lessening murk. Heavy rain succeeds, and if you're alive you go out and rescue the perishing.

The wind velocity in the path of a tornado is enormous,--anything up to 500 miles an hour,--but no instruments have been devised to withstand the strain. Varying pressures are responsible for the destruction. As the funnel passes over a house where the normal air pressure is about 2,000 pounds to the square foot it removes 1,500 pounds for an instant. Naturally the outside walls cannot withstand this enormous inside out pressure and the house explodes like a projectile. Only under such conditions could the vagaries of matter,--straws piercing logs and chickens bereft of every feather--be perhaps not explained but pardoned.

Stories of any degree of incredibility crop up after each tornado, often with accompanying photographs as proof. People are plastered with mud, pianos are deposited in neighboring lots, babies are hung up unhurt by their clothes in tree-tops, and often one person is killed and another nearby escapes unhurt, Bible-fashion.

Tornadoes may form almost anywhere, but they are never found on the immediate Pacific coast. They are most common in the Mississippi Valley, are rather common in the Gulf States, and have occurred throughout most of the East at one time or another.

Since there is no way of stopping them the next best thing is to know the conditions that make for their formation. If the Weather Bureau predicts a cold wave for sections of the country where the weather is already abnormally warm the line of meeting will probably produce a tornado somewhere. The officials, however, advise you not to worry until you see the intensely black cloud in the southwest trailing its funnel. See where this funnel is tending and run the other way. All tornadoes progress from the southwest to the northeast. Bad as they are, this makes them far less terrifying than if they whipped back and forth over a town or chased you around the pasture. If you happen to be in the house, take to the cellar, the southwest corner of it. If you can't escape lie face down to the ground.

The only tornado that I have ever witnessed was an undeveloped one in England, and a bit lethargic compared to those of the Prairie States. But even this blew an entire train off the track. It had all the other appurtenances of a tornado, the hail, the twisted trees, the narrow southwest to northeast path. The fact that the houses had only corners of their roofs blown off showed that as a tornado it was distinctly second-grade and without power to explode.

England, shortly after, was raided by three water-spouts. These phenomena are caused by precisely the same conditions as are the tornadoes. They form over the sea, and the funnel is composed of water. They take considerable bodies of water up into the skies and torrential rains result over adjacent districts. If I remember correctly, two of the English water-spouts broke against the cliffs and the other, moving inland in modified form, gave Gloucester a nine-inch rain. Ships have been known to fire cannon at these spouts. If one hit a boat directly damage might be caused, but they have little of the destructive force of the tornado.

As our country builds up the destruction from this most powerful of all phenomena is likely to increase. Bureau warnings over phones may result in the saving of some lives; cellars will undoubtedly be built in the principal zones. But the problem is an interesting one, for unlike the waterspout, cannon cannot be employed to shatter an emptiness that stalks the more malignantly the emptier it is.

THE HURRICANE

The tropical hurricane is undoubtedly nature's mightiest exhibit. The hurricane is the cyclone par excellence. It does not differ from our ordinary weekly cyclone in the essentials of wind rotation or pressures or rainfall; but it does differ in place of birth, in its course, and chiefly in its intensity.

The genuine hurricane is a West Indian production. It is generally cradled in those islands south and east of Jamaica and Cuba. It is nursed by the trade-winds. The first notice of its birth is an alteration in these winds, which are among the most regular observances on our planet. An extensive formation of cirrus clouds spreads over the sky and the barometer, which has been stationary for some days, edges off and begins a long and gradual fall. Great rollers are noticed for a day or two before the winds rise. A hurricane moves slowly.

This tropical organization is superior in depth to our shallow, disc-like, continental cyclone which is one and rarely over two miles thick. The hurricane rears its head three, four, and even five miles high. Instead, too, of dissipating its force over thousands of miles at once it is only a few hundred miles in diameter. Its center moves methodically along at the not very impressive speed of fifteen miles an hour, while our cyclones hurry along at thirty. But the hurricane is thorough. The wind about its center reaches a velocity of 120 miles an hour. This velocity has never yet been attained on the surface of the earth by our trans-continental cyclone.

Our cyclone always has an eastward trend; the hurricane has a parabolic course. It begins by moving west on the trades, drifting and dealing destruction to the banana and sugar plantations of Jamaica. It enters the Gulf of Mexico, and since it is then pretty much out of the influence of the trades it curves to the right and begins to act like any other storm by heading directly for the St. Lawrence. If it passes out through the Florida straits it never reaches the St. Lawrence but speeds up the coast and out to sea, usually at Hatteras to follow the shipping routes across the North Atlantic.

But if it has become so involved in the Gulf of Mexico that it cannot escape to sea again, it comes up through the Gulf States and on toward New England. Fortunately as it goes inland its intensity diminishes because it has not so much energy-giving moisture to draw from. Also its sphere of action widens, its embrace is less mighty, its characteristics more those of an ordinary continental cyclone. It manages, however, to deliver gales of 80 miles an hour along the coastal plain, increasing to 100 at the exposed places such as Hatteras and Block Island.

The intensest hours of a hurricane are those when its course is changing from westward to eastward. Enormous rainfalls accompany these storms, amounting to six inches in some instances. Since one inch of rain amounts to 100 tons per acre, and 64,000 tons to a square mile one can imagine the great amount of evaporation that has taken place to so saturate the air as to drench vast territories to such an extent.

While scarcely a year goes by without one of these West Indian hurricanes distinguishing itself on our shores the one that visited Galveston in 1904 eclipsed all. It chose to turn in the vicinity of the city. The gale increased to over 100 miles an hour and the wind gauge then blew away. The waters of the Bay were heaped up and three thousand lives were lost in the flood and wreck of flying houses. This peculiar storm did not turn northeast at once but ascended the Mississippi, turning at the Lakes and proceeding down the St. Lawrence after having spent a week in our country.

The listless doldrums have sent us 121 of these storms in the last generation. June has seen 8, July 5, August 28, September 40, and October 40.

Sea-yarners have seized upon the hurricane to energize many a flagging chapter, and particularly have they emphasized the eye of the storm. The eye is that vortex where contending winds neutralize each other into a calm, where the sun shines out through the scud, where the waves, relieved of the great pressure, leap upward in wild disorder. Then the center passes and the wind flings itself upon the unlucky bark from the opposite quarter. Its first onslaught is always represented as being the fiercest of the whole storm and gradually lessening as the center drives farther away. This is true in the same way that the first attack of the thunderstorm is usually the fiercest, both being when the pressure begins to rise. This savage change to the northwest is naturally the hardest of all for the ships to bear as they must steady at once against the severest blast instead of gradually bracing for its culmination. In no department of meteorology has fiction adhered so closely to the facts as in the sea-rover accounts of the hurricane.

But in real life there is very little excuse for the vessel to be caught anywhere near the disastrous center of the storm. Indeed, for generations sea-captains have known how to escape the deadly eye. By watching the barometer and noticing in which direction the wind is working round they can tell the course to a nicety and estimate its speed. Then the wise ones run the other way for even the Olympics and Imperators of the sea are cowed by the might of the West Indian.

The typhoons of the West Pacific are similar manifestations.

The hurricane moves off from its birthplace so slowly that our Weather Bureau has an opportunity to size it up, to chart its probable course, and to warn shipping interests. The ship-owners, as a class, appreciate the service of the Bureau and obey its warnings. Vessels with cargoes of a total value of $30,000,000 were known to have been detained in port on the Atlantic coast by the Bureau's warnings of a single hurricane. Now that a much vaster commerce will steam through these dangerous waters toward the Panama Canal the warnings will assume an even greater importance.

The best description of a hurricane that it has been my fortune to read is in a story entitled "Chita," one of the remarkable fictions of Lafcadio Hearn. As truthfully as a scientist and with great beauty of style he has pictured the long days of burning sun, the foreboding calm, the thickening haze, the ominous increasing swell of the ocean, a breathless night with the lightning glowing from between piling towers of cloud, the startling suddenness of the wind's attack, its fury, the hissing rain, the shrill crescendo of the gale.

CLOUDBURST

It is the American tendency to exaggerate. We call every snowstorm a blizzard, every breeze a gale, every shower a cloudburst. In our generous vocabulary it never rains but it pours. Consequently if we, in the East, ever had a real blizzard or a real cloudburst we should be at a considerable loss to find words for an unprofane description. I do not know how they manage out

West where these things occur.

A genuine cloudburst must be an amazing spectacle. It is caused by a furious updraft of wind keeping a rainstorm in suspense until so much water has accumulated that it has to let go all at once and the accumulation descends like a wet blanket.

This phenomenon is staged in the mountains; most often in the Rockies where melting snow and desert-hot ravines provide the necessary extremes of temperature. Wind blowing up a mountain-side can maintain considerable force,--so much that a man cannot possibly walk against it. Black thunder clouds brew on the peaks. Suddenly the collapse, and the person who tells the story afterward finds himself struggling in a torrent that a minute before had been a dry gulch. The moral of the story seems to be that if you are camping in the mountains and there is a strong upstream wind blowing and the clouds darken about the hill-tops and the thunder mumbles then don't make your bed in the creek-bottom lands. The high water marks of former freshets, but not of cloudbursts, show on the side of the stream.

Even in the less impulsive East a couple of inches of rain make a surprising rise in a little creek.

THE HALO

The halo is a luminous circle around the moon or the sun. It is caused by the refraction of light passing through moisture, which at the usual height is in the form of ice-crystals. The halo when complete consists of two large circles whose diameters are constant, 45 and 92 degrees. Then there are often other arches in contact. At each point of contact occurs a parhelion which is a mock sun of brilliant colors and called a sun-dog. Since the sun-dog is brighter than the other parts of the halo it sometimes appears when the rest of the halo cannot be seen. Sun-dogs hunt in pairs or fours. If the halo is colored the red is on the inside. When the colors are caused by diffraction instead of refraction, the red is on the outside of the prismatic ring and the halo is called a corona.

Having now satisfied the demands of science all that can be forgotten except that the halo around either sun or moon means excess moisture in the atmosphere. The wide halos are seen in the high cirrus clouds 25, 36, 48 hours in advance of a cyclone. At first the ring is very wide and faint with several stars in it. If the storm is advancing rapidly the halo brightens and narrows and the stars fade. This is proof to show that the proverb stating that the number of stars inside the ring is a forecast of the number of days of storm is sheer nonsense. For presently the ring closes and the stars disappear which would show according to the proverb that the storm had changed its mind and would cut down the number of days from several to none.

The moon grows paler. The light that it casts upon the earth is eerie at this stage. Within a few hours the cocoon of mist is completely woven about the moon. The circle has closed. Snow or rain begins within a few hours after the moon has entirely disappeared. If it does not so begin it shows that the process of increasing humidity is a very slow one and the storm center is probably passing far to one side of the observer. Also if the snow begins before the light of the moon is entirely suppressed the disturbance is a shallow one and the storm will be light.

When the halo is actually a corona (red outside) the approach of the storm can be gauged by the rapidity with which the circle grows smaller. For a decrease in diameter denotes that the size of the moisture drops is increasing and therefore the storm is approaching. As a matter of fact the corona will have disappeared long before the time for rain. Still it is useful to know that if the corona increases in size the conditions are clearing. With the halo the reverse holds. For when the clouds are very high the halo looks small, and high clouds imply swifter winds and a greater distance from the storm center.

The Zulu Indians who have an eye for the picturesque as well as for the truth state the chief fact about haloes happily: "When the sun is in his house it will rain soon." Another saying of theirs anent cumulus clouds holds for our country as well as for theirs: "When the clouds rise in terraces of white, soon will the country of the corn-priests be pierced with the arrows of rain."

There are many little observations which the man who has kept the corner of his eye open may profit by and yet which are rather difficult to express in type. Who could describe an egg for instance whose springtide of youth was far behind and yet was not quite ready for the discard! In nature it is the fleeting moment of transition, the half-tones of the border that are so hard to catch, so difficult to portray, and yet so very important not to miss if one is to become sure. There follow some of the baldest and most communicable half-facts about the weather that should be used oftener to bolster up some opinion gleaned from more positive sources than to mould one in their own strength.

Moisture in the atmosphere helps sight to a certain extent. For when the air is full of moisture its temperature tends to become equalized, obliterating irregularities which would otherwise reflect the vibrations producing sight and sound. So if one hears better or sees better on a certain day it augurs a moister atmosphere,--an auxiliary sign if there is a view that you are fond of looking at many times a day. In the city, alas, clearer vision on one day than another means merely that less coal is being used. But in camp there is very often a perceptible difference in one's seeing ability even on days that could all be classed as clear.

Another thing that the haunter of the woods may notice is that his smelling capacity is increased before a storm. The increase of humidity which precedes a rain buoys up odors and depresses smoke. Even in dry weather if you will stroll by a marsh you will notice how rank the vegetation smells and how the smells float in layers in the air strata of different humidity. One's sense of smell is a very slender thread on which to hang a storm, however.

Fires burn more briskly in dry air than in moist, but to tell the difference (if you can't feel it) you must be very sure that your wood is as dry on one day as on another.

Before a rain many plants close their flowers or shift their leaves. The dandelion, pimpernel, red clover, silver maple are good examples of this, but they would not be of much use in the North Woods. The closing, too, takes place only a few hours before rain and is merely confirmation of the signals rendered more adequately by clouds and winds.

Bugs and flies are particularly annoying before a storm and it is surprising that the spider should not take advantage of this to get a meal. But spiders are cautious and they never spin a web on the grass, at least on the day that brings a storm. The insects do not fly so high on these weather-breeding days and consequently the birds that feed on them fly lower. The chimney swifts are a particularly good guide to the different altitudes at which insects fly.

The stars are on a par with bugs as weather guides, although there are many proverbs that grant them much. One circumstance should not be neglected, however, and that is that wind mixes air and when air is well mixed atmospheric inequalities are less disturbing to vision. Hence when one can see the stars and the moon well wind currents are oftenest the cause. Even if it is not blowing on earth these wind currents may yet be blowing above to reach the earth later. In this way cold waves arrive. There is an old proverb about this condition, applying it to the moon, "Sharp horns do threaten windy weather."

But the stars are of second rate importance because they are so soon obscured. If you can't see them it is cloudy, but you do not know what kind of cloud it is. If only the brightest show, a veil of cirrus is arriving. A dark sky with only a few dim stars is an omen of storms. If the stars twinkle it is because the varying currents of the upper air are in juxtaposition. If they twinkle while the northwest wind is on it is a sign of colder weather,--not because they are twinkling but because of the northwest wind.

In the days when almanacs were the sole guides to the weather a man with a sense of humor, Butler by name, got out one and dedicated it to "Torpid Liver and Inflammatory Rheumatism, the Most Insistent Weather Prophets Known to Suffering Mortals." Rheumatism is following the almanac to the scrap heap, and it would be harder for a camper to guess what a torpid liver was like than to forecast the weather, yet for the majority of "suffering mortals" there is still much truth in the amiable observation of Mr. Butler,

"As old sinners have old points O' the compass in their bones and joints."

CHAPTER V

THE BAROMETER

Whatever the foregoing chapters may imply as to the whole world going camping the fact is that the woods are still, unfortunately, for the few. The woodsman must yield gracefully to the suburbanite,--in numbers.

But the weather is for everybody. To be sure the sunrise that talks so confidentially to the hunter of the coming day does not exist for the commuter. But the coming day does, even though the things it means are essentially different. To the hunter with his seasoned clothes and well-earned health a rain is only of concern in so much as it affects the business of the day; personally it is of small moment. But to the commuter what does the weather mean? Dollars and cents, of course. His business goes on, but to his person one unexpected shower = the cost of pressing a suit; one thorough soaking = one doctor's bill. For you cannot expect the man to throw off a chill who can quiet his conscience on the matter of daily exercise by watering the geraniums and reading the newspaper.

Weather wisdom is necessary for the hunter; for the commuter it pays.

The hunter had to rely on local weather signs. The commuter can go him one better by investing $10 (how finance will creep in!) in a little aneroid barometer. The local weather signs were good for twelve hours at the longest. The barometer is a faithful instrument that adds another twelve hours to a man's knowledge. Half a day, or even a day before any local sign of changing wind or growing cloud appears the barometer is on the job. It will register in Philadelphia the news of a disturbance approaching the Mississippi. So sensitive is it that it is the slave to every wave of the great air ocean.

The barometer gauges for the eye the amount of atmosphere that is piled above one. If the amount is normal and at sea-level the instrument will measure 30.00 inches. This air pressure is equivalent to a column of water 30 feet high. As this would make unwieldy prognosticators the scientists use mercury instead, which requires a column less than three feet long. And for general purposes this is supplanted by the handy little aneroid (which means "without fluid"). This is so fixed that the pressure of the air influences the upper surface of a vacuum chamber, balanced perfectly between this pressure and a main spring. This action is transmitted to an index hand moving across the dial marked into fractions of inches after the manner of the recognized standard, the mercurial barometer.

When the warm moist light air of a cyclone invades a locality the pressure is partially removed, the vacuum chamber is not pressed so hard and the dial hand or the mercury subsides. When the cold, dry, heavy air of the anticyclone lumbers in more pressure is applied and the mercury, or the dial hand, climbs. So a falling barometer means a storm, a rising one fair weather.

That is a generality that glitters. If that were all there was to it weather officials would have a sinecure. But each cyclone varies in size, intensity, and rate of progress. Some do not advance for days. Therefore there has grown up a pretty large body of information as each storm has had to be watched and the barometric movements recorded. The most important variations follow:

Remembering that 30.00 inches is sea-level normal, if the barometer is steady at 30.10 or 30.20 the weather will remain fair as long as the steadiness continues, and on the turn, if the fall proceeds slowly with the wind from a westerly direction fair to partly cloudy weather with slowly rising temperature will follow for two days.

If the barometer rises rapidly from 30.10 the fall will be equally rapid and rain or snow may be expected within a couple of days. Since the depressions of the atmosphere tend to a certain regularity about the center of the storm it follows that the reactions will follow the actions in similar manner,--a long rise portending a long fall and a variable glass meaning unsettled conditions.

The barometer does not rise with wind from an easterly direction unless a shift is imminent. In winter the air is so much colder over the land than over the sea that the air brought in by an easterly wind is soon condensed. Consequently with winds from the south or southeast, even if the barometer is 30.20 or 30.10 and falling slowly rain usually arrives (and rain of course is meant to include snow whenever the mercury is below the freezing point) within 24 hours. If the fall is rapid there may be precipitation within 12 hours, and the wind will rapidly increase and the temperature rise.

If the wind is from the east or northeast and the barometer 30.10 or above and falling slowly it means rain within 24 hours in winter. In summer if the wind is light rain may not fall for a day or so. If the fall is rapid in winter rain with increasing winds will often set in when the barometer begins its fall and the wind gets to a point a little east of north.

If the barometer is 30.00 or below and falling slowly with northeast to southeast winds the storm will continue 24 to 48 hours. If the barometer falls rapidly the wind will be high with rain and the change to rising barometer with clearing and colder will probably come within 20 to 30 hours.

If the barometer is below 30.00 but rising slowly the clear weather will last several days.

If the barometer is 29.80 or below and falling rapidly with winds south of east a severe storm is at hand to be followed within 24 hours by clearing and colder. Under the same conditions but with northeast winds there will occur heavy snow followed by a cold wave.

If these promises do not always bear fruit it is because they will have been interrupted by an unseen shifting of the atmospheric weights. But the barometer will record them. A rapid rise may be checked in ascent and the instrument may fluctuate like a stock-ticker. Its tale is of very unsettled weather conditions and consequently no particular brand of weather will last for very long at a time.

A sudden rise of the barometer may bring its gale of wind as well as a sudden fall. But the tendency will be toward clearing and much colder.

A fall of the barometer on a west wind is not common. It means rain. A rise on a south wind means fair. A low barometer and a cold south wind mean a change to west with squalls for a while. On the other hand, a high barometer with warmer weather means a shift of the wind to southerly quarters and an imminent fall.

If the barometer rises fast and the temperature does, too, look for another storm. This is often noticed in summer.

There is a slight daily oscillation of the mercury, which, if other things are steady, registers highest at 10 A. M. and 10 P. M. and lowest at 4 A. M. and 4 P. M.

If this data confuses bear in mind the simple ordinary progress of the barometer in the usual storm: First, it will stand steady for a day or so at any point between 30.10 and 30.50. Then the glass will begin (for most storms) to fall gradually. As the center nears the fall hastens. After the lowest point has been reached a slight rise will be followed by another slight fall and then the final long rise will commence. The rain begins and ceases at different stages for different storms, depending upon the wind's velocity and direction.

For every 900 feet of altitude the height of the mercury is about one inch less. Do not complain that your barometer is inaccurate if you are living up in the mountains and your readings are not the same as the weather reports which are reduced to sea level. All the figures given in this chapter are for sea level and if your house is 1900 feet above you must move the copper hand of your aneroid 1.95 inches from the pressure hand. If the pressure hand would read 28.05 the adjustable copper hand would read 30.00 which is the sea level reading.

One good thing to remember is that a barometer falls lower for high winds than for heavy rain. A fall of two- or three-tenths of an inch in four hours brings a gale. In the ordinary gale the wind blows hardest when the barometer begins its rise from a very low point.

In summer a suddenly falling barometer foretells a thunderstorm, and if the corresponding rise does not at once take place the unsettled conditions will continue with probably another thunderstorm. If you see the thunderstorm first, that is, if the barometer is not affected by the approaching black cloud you may be sure that the storm will amount to nothing.

The man in the fields or along the shore has many natural barometers in animal life. But these natural barometers only corroborate; they do not foretell, at least very long before. Some are useful at times and among these the birds are foremost. The observant Zulus have incorporated this in one of their pretty proverbs, "When chimney swallows circle and call they speak of rain." As a matter of fact the swallows are circling most of the time after insects. If they are flying high it is because the bugs are flying high and that is because there is no danger of rain. As the rain nears the air gets moister, the bugs and the birds fly lower.

Whether they do this because their instinct is to avoid a wetting or because the lighter atmosphere of a cyclone makes flying more difficult, particularly at altitudes, I do not know. For weather purposes it is enough to watch their comparative levels. Wild geese are excellent signs, I am told, but it would be a dry country that waits for a sight of them for its rain.

Bees localize before a storm and will not swarm. Flies crowd upon the screens of houses when humidity is high, possibly because the appetizing odors from within are buoyed afar by the heavy air. Cuckoos seek the higher ground in fair weather and disappear into bottom lands before a rain. Although they are called rain-crows they are heard in all weathers.

Smoke is as good an evidence of barometric pressure as anything except the instrument itself. On clear, still days it will mount; on humid days without wind it will cling to the hill. There is that difference. But it takes skill and many comparisons to gauge its angles in the wind. It becomes a test in observation and finally rewards one by becoming an excellent sign not only of air texture but of the direction of its currents.

No reference to barometers would be complete without mentioning spiders. They show a most delicate apprehension of changing conditions. If the day is to be fine and without wind they will run out long threads and be rather active. If the rain is nearing they strengthen their webs, shorten the filaments and sit dully in the center. Fresh webs on the lawn insure a clear day. But for the commuter, whose time is money, there is little leisure to consider the spider.

As a natural result of the variation in altitude affecting the barometer the words which are printed on the face become entirely useless. In some places it would be impossible for the needle to point higher than "Very Stormy." Even at sea level a sudden fall to "Fair" would cause a rain, much to the indignation of the person who thought that he had purchased a self-registering weather prophet. Disregard the words but watch the needle and you will never be surprised at what the weather is doing next.

CHAPTER VI

THE SEASONS

Too great emphasis cannot be laid upon the futility, at present, of trying to forecast the weather for more than a very few days in advance. Long range efforts are not made by the Bureau because with its present limited knowledge of the factors that control seasons and with the present limited facilities for collecting data the process of looking into next month has not been perfected, and the attempt to investigate next winter's weather proves scientifically impossible.

As usual, fakers step in where science fears to tread. With goose-bones (not their own) and hickory nuts they prophesy with all their might. And if their prophecies come true, as sometimes they must, there is wide rejoicing in the newspapers and the cause of science is set back by just so much. But science cannot be thwarted in the end and every year new discoveries are made, new speculations proved true or forever false, and some time, doubtless, the weather will be predicted from year to year with the same 85% accuracy with which the 36 hour forecast is now made. Experimenting is worth the little that it costs, too, for to know when the summer is to be dry or wet, hot or cold will be a boon to everybody and to the farmer most of all.

One conclusion has already been reached by officials in the Weather Bureau and scientists generally. It has been decided by long search through creditable records, painstaking comparisons of averages coupled with the most accurate investigations for half a century, that, on the basis of ten years, our seasons do not change. That is, counting the decade as a unit, our weather keeps to the same level of efficiency through the centuries.

This statement comes always as a blow. It always provokes argument and citations of grandmother's blizzards. There is a great and universal hesitation in believing that our weather is as good to-day as it used to be. The good old times when there was a general debauch of snow and you could skate all winter on anything but the Atlantic Ocean certainly appear no more. As a matter of fact there has been a change, but it has been in our memories. In grandmother's youth the trains,--if they had trains then,--doubtless were stalled by a big snow for then they did not have rotary plows. In father's day they may have had an unbroken winter of sleighing. We couldn't now; sleighs are extinct. But in our time, in fact every year, some record is being broken and the records go back a respectable length of time.

For example in Philadelphia the most accurate records made by standard instruments have been kept for 43 years. During this time the highest wind velocity was recorded in 1878 (75 miles an hour). The greatest rainfall in 24 hours occurred in 1898 (5.89 inches). The lowest temperature was registered in 1899 (6 degrees below zero); the highest in 1901 (103 degrees). The greatest number of thunderstorms for any one year took place in 1905 when we had 51. As late as 1909 the heaviest snowfall ever recorded at this station, amounting to 21 inches, occurred. And just a few weeks ago (April 3rd, 1915) it snowed 19 inches in half as many hours. All these items do not indicate a climate decreasing in virility very swiftly.

But there is more evidence yet that Philadelphia is experiencing the same varieties of weather in about the same proportions. Diaries of observant men running back to 1700 show that almost any kind of memory could be founded on fact, that the same violent changes in temperature, the same deep snows and unseasonable seasons that we endure to-day were noticed then. To quote:

"The whole winter of 1780 was intensely cold. The Delaware was closed from the 1st of December to the 14th of March. The ice was from two to three feet thick." We despaired of ever living up to this until three years ago when the same thing happened and sleighs crossed the river a little above the city. And despite the new ice-boats!

"The winter of 1779 was very mild, particularly the month of February when trees were in blossom."

"On the 31st of December, 1764, the Delaware was frozen completely over in one night, and the weather continued cold until the 28th of March with snow about two and a half feet deep."

"The winter of 1756 was very mild. The first snow was as late as the 18th of March."

And so it goes. 1750 was mild; 1742 "one of the coldest since the settlement of the country"; 1741 was intensely cold, 1725 mild, 1714 very mild after the 15th of January, 1697 long, stormy and severely cold. The upshot of it all is that February violets and April snows were just as well known to General Washington as they are to us.

Nimbus is any cloud from which rain is falling, and the important thing to know is how to judge from the formless thing how much longer it is to rain. The wind is the surest guide. In this picture the nimbus cloud is only that at the end of the cape. All the rest is torn stratus and cumulus, which needs to condense a little further before it becomes nimbus. This will likely happen because the cloud at the left is very dark. The broken appearance denotes some wind. Rain does not fall from a mottled sky nor yet a streaky one; the nimbus is uniform in appearance. In summer a break in the nimbus will show a veil of cirro-stratus above. Just nimbus by itself will not support much of a storm. In winter if the nimbus is particularly seamless snow is about to fall.]

But though all facts point to the fact that the climate does not change in a decade or a generation or a dozen generations, there is some comfort for those who are not satisfied in knowing that it doesn't stay the same forever. During the carboniferous times the poles were as warm as the tropics and when the Ice Age came on it was very chilly everywhere. If one might only live an eon or two he might then well complain of the changing climate.

Climate, however, is one thing, weather another. The climate is the sum total of the weather. Climate is as enduring as our Constitution, the weather is as changeable as our city governments. No matter how proud a scientist may be of the lasting qualities of the climate, he has to admit that our weather, taken day by day or even year by year, is versatile in the extreme. And the question he has set himself to solve is how to explain the variations of the seasonable weather. He wants to find out why all winters are not alike, and why no two successive springs are the same. Then he will be on firm ground at last and able to make scientific forecasts for the ensuing year.

The obvious thing was to find out as accurately as possible what had happened and science's keenest eye was focused on records in the hope of discovering fixed periods of warmth or wetness, cycles of cold and drought. So far no cycles have been discovered that are beyond dispute. Nothing has been found that cannot be contradicted successfully. This is discouraging.

One of the most frequent starting places for investigators is the spots on the sun. They found that periods of three, eight, eleven, and thirty-five years should bear some resemblance; 1901 was eagerly looked forward to. They wanted it to correspond with the remarkably cool summer of 1867. When it started off in July with a temperature of 103 degrees, the highest ever recorded in Philadelphia, they concluded that the sunspots were fooling them. A connection between sunspots and weather has not been established, therefore, although they are now known to affect the electrical condition of the earth's atmosphere. Longer periods of observation will permit comparisons that may yet define concurrent cycles of sunspots and weather.

A definite weather cycle has not yet been discovered, but one step in the way has been cleared up. We now are pretty sure of one cause for unusual single seasons of heat and cold.

There exist in winter great bodies of cold, dry air heaped up over Canada and Siberia, which are formed by the greater rapidity of radiation over land surfaces than over water. These mounds of cold air build up during December, January, and February and form great so-called permanent areas of high barometer. It is on the skirts of the Canadian high that the smaller highs form which sweep over our country, giving us our cold waves. Also in winter permanent lows form over the North Pacific and North Atlantic where warm currents afford continuous supplies of warm moist air. From the great Aleutian (Pacific) low spring most of the cyclones which swing down below the border of the Canadian high, make their turn somewhere in the Mississippi Valley, and then head for the Icelandic low.

It can be seen that if the Canadian high is a little stronger than usual and spreads a little farther south, then the northern half of our country will come more directly under its influence and we will experience an unusually severe winter. As the storms are pushed south and as the cold air pours into the northern quadrants the snow line is pushed south too. Hence all abnormally snowy winters are caused by a strengthening of the permanent Canadian high which may be central anywhere north of our Dakota or Montana borders.

Conversely, if this high is weaker than usual the cyclones can cross the country on a line farther north, there will be less snow, and the cold waves that follow will be less severe or even non-existent.

In summer the reverse occurs. Great oceanic highs are built up over the South Atlantic and South Pacific and a permanent low occupies the center of our continent. The character of the season is determined by the strength and position of these areas. The eastern states are affected especially by the slow movements of the South Atlantic low. The puzzle is why should these areas change their power and position, and if they must change why don't they do it regularly? The puzzle will undoubtedly be solved. These great centers of action will be plotted against and observed from every vantage point by a thousand observers. A fascinating field for scientific speculation opens.

At present our Government exchanges daily observations with stations in Siberia, Canada, and the West Indies. The great storm-breeder, the Aleutian Low, is watched from Alaskan shores. In the Atlantic the Bureau needs stationary ships to record the growth and decline of the High over the Azores. Knowledge of the wind circulation from this would inform us whether our storms were to be shunted farther north and pushed somewhat inland. A storm which is pushed to the left of its normal track increases tremendously in intensity. Whereas a cyclone that limps slackly to the right of its normal line loses intensity at once. It misses coil. In this respect storms seem to resemble rattlesnakes.

The energy of the Azores High influences the number and destructiveness of the West Indian hurricanes: the larger the area is the closer do the hurricanes hug our shores and the more destruction do they accomplish.

The very sureness that the general average of the seasons is to be the same enables us to guess pretty accurately for individual purposes as to the kind of season coming next. A guess, let me add, is not a forecast. It is a gamble and disapproved of by the Bureau, but until they supply us with a basis for judgment we will have to go on guessing, for human curiosity is as near to perpetual motion as the weather is to the lacking fourth dimension.

One of these guesses is that if the winter has been a warm one the summer will be cool, for the very good reason that the yearly average does depart so slightly from the fixture. Unfortunately one hot summer does not mean that the following summer will be cool. Certain sequences of the seasons have been observed often enough to have been gathered into proverbs. Everybody agrees that "A late spring never deceives." "A year of snow, Fruit will grow." "A green winter makes a full churchyard."

Of the many hundreds of proverbs relating to the seasons a few are sage, some outworn, and many sheer nonsense. Nearly all refer to the obvious fact that one kind of season is followed by another rather unlike it, not much telling what. And there, unsatisfactorily enough, they leave one. But much is to be hoped for from the scientific explorations now in progress. And until they are heard from few of us will realize how many seasonable seasons we really enjoy.

CHAPTER VII

THE WEATHER BUREAU

At the cost of a cent and a half a year apiece we Americans are supplied with detailed information in advance about the weather. And the information is correct for more than four-fifths of the time. If stock brokers never missed oftener, what reputations would accrue!

Cheapness, accuracy, and a certain modesty are the three qualities that distinguish the out-givings of the Bureau from the old-fashioned predictions of the weather which used to appear in almanacs. Almanacs have probably kept appearing ever since the art of printing first allowed unscrupulous persons to juggle with words. They cost fifty cents and their predictions were based on nothing but the strength of their author's imagination. Of course, it was impossible for him to guess wrong more than half the time so that when he announced in January that July would be hot with thunderstorms he was often right. This gave him prestige, but aided his clients little.

The Weather Bureau was in about the same position in regard to the quack predictions of the almanacs as was the honest doctor of the last decade who could only prescribe good food and fresh air and moderate exercise for the patient who much preferred the expensive allurements of the medicinal cure-all as advertised. In humility the Bureau said that as things stood it could not forecast with accuracy for more than 48 hours, and its honesty brought it into disregard.

But, although the Weather Bureau,--like the Christian Church and other things that have had to combat superstition at every step--has grown slowly it has grown surely and its work is being recognized more widely and relied upon more understandingly every month. It was an American scientist who discovered the rotary motion of cyclones and their progressive character, but due to the conservative nature of our Government three other nations had established weather services before we had. In 1870 the War Department was authorized to start a system of observations that would permit of a rough sort of forecasting. The forecasts proved of so much value to shippers and sailors that the work was handed over to the Department of Agriculture and enlarged (1891). To-day every part of our country contributes to the knowledge of existing weather conditions.

At 8 A. M. observations are made at hundreds of stations and wired to the Central Office at Washington. The Chief there, knowing these conditions, is enabled to locate a storm, to gauge its rate of speed, to learn its course, and to measure its intensity. He can dictate storm warnings and be sure that within an hour every sailing master will have a copy. He can detect a cold wave at its entrance into our territory and know that within an hour every shipper, every truckster (who has signified that he wishes to be informed) will have the facts that will save him money.

At 8 P. M. the same stations telegraph the changed conditions, and if any very violent disturbance is in progress an observation is made at noon. Besides the Washington distributing station there are 1700 others from which warnings are sent by telegraph, telephone, or mail. There are 100,000 addresses on the mailing list and 5,000,000 telephone subscribers can get them within an hour. The newspapers reach many millions. And all this at a cost of 1-1/2 cents a year. If we, in a fit of generosity, should pay 2 cents, or even 2-1/2 the Government would be enabled to work out many of the larger problems awaiting only a larger appropriation to be attacked.

The people's investment of $1,600,000 a year is a good investment. In one year the Service saves a great many hundred per cent. A few known savings are worth giving; $3,500,000 worth of protection was made possible from one exceptionally severe cold wave; the California citrus growers estimated that one warning saved $14,000,000 worth of fruit; $30,000,000 of shipping (and cargoes) was known to have been detained in port just on account of one hurricane warning, and there are many warnings of gales every year. Uncalculated savings have been effected among the growers of tobacco, sugar, cranberries, truck. The railway and transportation companies save, through use of the forecasts, in shipments of bananas, oysters, fish, and eggs.

Farmers, manufacturers, raisin driers, photographers, insurance companies, and about a hundred and fifty other occupations increase their profits by a systematic study of the forecasts.

The people who live along the rivers often owe their lives and frequently much of their property to telephone warnings of approaching floods. The flood stages in all the principal rivers and streams have been calculated and losses are reduced by 75 per cent. by accurate predictions as to when the crest of the flood may be expected and how high it will reach. A hundred uses of river forecasts, even when flood stages are not expected are given in the booklet, "The Weather Bureau" which you can have from Washington for the asking, like many another of their publications.

Yet, with all the good it does, the man on the street still regards the Bureau as an uninteresting, undependable exhibit in the upper corner of the newspaper,--if he regards it at all. It is his child, however, who is instructing him. For his child is being taught in the public school all about it and he takes his teaching home and becomes the teacher. The child is father of the (old) man in lots of instances.

The most impressive thing about the whole output of the Bureau to the child is its Map. The Bureau issues a map every day which is posted in post-offices and railroad stations and in schools, too, if they ask for it. And every day this map shows in all its gripping details the way our storms are sidling across the continent or rushing up our coasts. It prints the word low where the stormy area of low barometer is. About the low run continuous black lines numbered 29.7, 29.8, 29.9, etc., which show where in the country the pressures are the same.

As the numbers run up to 30.0, 30.1, 30.2 they begin to circle about the word High which denotes where the pressure is highest. Little circles will be observed on the map. Some are clear, indicating clear weather; others are half clear, half black, indicating partly cloudy conditions; others are all black, showing clouds; others have R. or S. inside them, telling where it is raining. The numbers under the circles show how much it has rained or snowed and the numbers under the other numbers are the velocities of the winds. The arrows through the circles fly with the wind. A little zig-zag locates each thunderstorm and the shaded portions show over what portions of the country it has rained during the last 24 hours. As an intelligent puzzle picture the map is unequaled and no wonder the child likes it.

With this map you can tell at a glance what the weather is doing to your uncle in Tacoma and to your cousin in Missouri. With two successive maps you can find out about how fast the storms are traveling, in what direction, and how low the temperatures are under their influence, and so estimate for yourself the weather for the next three days.

Besides the invaluable daily weather map the Bureau issues many other maps that present the phenomena of the week, the month, and the season in graphic form. Masters of vessels are now co 鮨 erating with the government to provide observations at sea, and both on our northwest and southeast coasts such information is very valuable. In the west several hundreds of stations are maintained in the mountains for the purpose of obtaining the depth and content of the great snowfalls there. Estimates can then be given out as to the amount of water to be available for irrigating purposes. In addition to the 220 stations of the first class there are 4200 co 鮨 erative stations at which observations are made and mailed to 44 centers for distribution.

Special local data help to establish the relations between climate and forestry, agriculture, water resources, and allied subjects. Many bulletins are compiled by experts in their respective lines and these are for free distribution. A study of forest cover is being made in Colorado and the effects of denudation on the flow of streams will soon be scientifically established. As soon as practicable the Bureau hopes to extend its period of forecasting. Weekly forecasts have been tried in a general way with success, but long-range forecasting depends upon so many relationships of the air that present knowledge and facilities do not warrant its adoption.

CHAPTER VIII

A CHAPTER OF EXPLOSIONS

In the good old times when a man was born, spent his life, and died in the same village the weather proverb was fashioned. Generations had watched the clouds gather under certain circumstances and scatter under certain others and they naturally drew conclusions. These conclusions crystallized until they resembled nuggets of golden weather wisdom. Some were even used as charms. And all contained a deal of truth so long as they were only meant to refer to the country in which they had originated.

But nowadays when the very idea of remaining in the same place for very long at a time is obnoxious the weather proverb suffers. It suffers chiefly by transportation. The weather in County Cork is so very different from the weather that makes Chicago famous that the same weather lore does not fit. Yet it is often applied. The old truths, treasured in picturesque phrase and jingle, were brought over the ocean unchanged and made to do duty,--a case of new wine in old bottles again, for a gentle old Irish proverb splits up the back when it tries to accommodate itself to a week of our reckless but magnificent weather.

Fairy stories are jewels to be cherished. And it is a careless and unimaginative race that perpetuates no legends. Even old saws are quaint and should be preserved: "See a pin and pick it up, all the day you'll have good luck." Let that sort of thing go on because it adds richness to our conversation. But if a thousand men, after having picked up their morning pins, sat around waiting for the ensuing luck the progress of scientific business management would be halted. And precisely that way is the knowledge of ordinary weather facts halted,--a full-grown superstition sits in the path. Instead of relying upon their eyes the majority of people rely upon a bit of doggerel. For example, millions of people firmly believe that the ground-hog is a key to the weather. They say that if the ground-hog does not see his shadow on the 2nd of February that winter is over!

This is the sort of thing that obscures the findings of science not to mention common-sense. Few of these people have ever seen a ground-hog. Few of the rest have ever studied its habits. The ant, the mouse, the fly, the rat, and the mosquito have far more influence upon our lives than the ground-hog has and the most ambitious animal cannot expect to influence atmospheric pressure, which is responsible for our weather. Yet as often as the 2nd of February comes around the hopes of many are either dashed or raised according to the actions of this creature. As a matter of fact, whether February 2nd is clear or cloudy can have no influence on the rest of the winter.

Almost all the other proverbs have a basis of reason. But this puts its believers in the wrong either way. If they say that it is the actions of the animal that they rely upon they depend upon a characteristic thoroughly and surely disproved. No animal, although it may sense a change in the weather a few hours in advance, is able to feel it for three days ahead to say nothing of six weeks. If these people say, on the other hand, that a cloudy February 2nd means an immediate and complete let up of winter, or that a clear February 2nd means a certain continuance of cold weather for six weeks, they have only to trouble themselves to look at the files of the nearest Weather Bureau for the last forty years. They will find no connection. The trouble is that they will not look, but keep on repeating the bit of nonsense and believing in it, although the strength of their convictions probably does not reduce their coal-bills.

The same people are fond of saying that the first three days of December show what the winter will be like. That is, if the 1st is fair so will December be; if the 2nd is cold so will January be; and if it snows on the 3rd, so will it snow in February. If all three should be clear and warm certainly a remarkable winter would follow! No rain, no snow, no cold! You see how absurd this superstition is.

"A dry moon lies on its back!" After the ground-hog the moon is supposed to have the most influence on our seasons. The Government and many scientists connected with no governments have made careful, exhaustive and conclusive investigations. No relation between the moon and our weather has been discovered except as she causes our tides and they affect atmospheric pressure in an infinitesimal degree. We would still have just as much and just as variable weather if there were no moon. The weather changes with the changing moon, and it does not change as the moon changes, and the chances are about even that the times of change will coincide. So there is, therefore, absolutely no foundation for the dozens of proverbs that yoke the changes of the moon with the changes of our weather. Neither in science nor in observation has any sequence been deduced.

So the moon may lie on its back or on its side or stand on its head and the weather will remain dry if no low pressure areas cross the country, and it can lie on its back for days and the country be drowned out if they do. There are enough pretty things to say about the moon, anyway, and will be more all the time for, to commit a paraphrase: Science is stranger than superstition.

"It will rain for forty days straight if it rains on St. Swithin's Day," which, I might as well say for the benefit of those who don't know their saints, falls on July 15th every year. It would be interesting to know how many people in a hundred really believe this, or really believe all the other things that are attributed to the saints,--quite a few, probably. Luckily for St. Swithin July and August are wet months, with often several days of showers or thunderstorms in succession. But never once in Philadelphia has it rained for forty days, one right after another, although half the July 15ths have been rained on. This proverb is one of those that had better never been transplanted from its native Ireland where rain for 40 days would excite scarcely a curse.

"Long and loud singing of robins denotes rain." It does not. Oftener than not it denotes the time of day. Just watch the robins and listen to them and see what they do before a storm, during it, after it, and then you will see how little the songs of birds can be depended upon to supplant the barometer.

"If March comes in like a lion it will go out like a lamb," and the other way round. I have seen March come in like a lion and go out like a lion, come in like a lion and go out like a lamb, come in like a lamb and go out like a lion, and come in like a lamb and go out like a Noah's ark. But I never have seen March do anything dependable. It is quite impossible to tell how March is going out on March 27th, and absolutely impossible to tell on March 1st.

But there is this much observation expressed in the proverb, that March is so changeable that, if it comes in cold, windy, unsettled, there is not so much chance for such weather still to be going on at the end of the month, and still less in England where the proverb came from. This is a harmless proverb unless it should lead people to actually count upon a pleasant spring just because March had an unpleasant inception. Misfortunes rarely come singly, even on the weather calendar.

"When squirrels are scarce in autumn the winter will be severe." Aside from the scientific truth that the animals cannot know in advance about the seasons there is little evidence on either side to base a contention. Nobody has made a squirrel census; nobody, probably, has found out whether they increase in numbers for six years and then die off in great quantities as do the rabbits in the north country on the seventh; nobody has connected their apparent numbers year after year with the actual severities of the winters. And so nobody has a right to promulgate the report (except as a bit of nonsense like April Fool) that the ensuing winter is going to be a record breaker because the squirrels have disappeared. It would be far truer to say that "When squirrels are scarce in autumn the hunters have been busy," and let it go at that.

There are a lot of proverbs in this connection about goose bones and hickory nuts and wild geese, which sound plausible but are never proved. If the birds have all the sense credited to them it is strange that some allow themselves to be caught by an early snowstorm in the fall and decimated. Also it is not uncommon for early migrations in the spring to arrive in the north to be slain by the thousand by a belated blizzard. It is granted that animals and birds, having a far greater sensitiveness than man, occasionally sense a catastrophe some hours before it is evidenced by any visual signs, but seasonal wisdom has not been proved in any one instance and disproved in many. None of the proverbs relating to the animals and birds are to be depended upon. They deceive, much to the regret of all the meteorologists who would welcome any genuine clue to nature of the coming season. Any farmer would be only too glad to keep a menagerie of squirrels and wild geese and toads if only he might be assured by them of the coming seasonal conditions.

The proverbs given indicate the range, possibly, but certainly not the full absurdity of the old weather sayings. There are many other proverbs that contain at least a half truth.

"Enough blue sky to make a Dutchman's breeches indicates clearing," is one that is true if the wind has changed to the west. If the wind still blows from an easterly quarter blue sky for a Dutchman's whole wardrobe would not insure clear weather. All sayings must be tested many times before they are believed implicitly.

"There is always a thaw in January," is about as true a generalization as can be made about things for which generalizations are never strictly in place. Even in Canada the severity of the winter is often broken by a spell of warmer weather with a rain, perhaps, in the dead of winter. In the United States a winter without some break in each of the months would be a most unusual occurrence. So that it is quite reasonable to expect the "January thaw" any time from Christmas until the middle of February.

"A late spring never deceives," unless it is so very late, like the phenomenal spring of 1907, that the jump is made, perforce, into summer. That is a cruel deception. What is meant of course is that if the freezing weather continues consistently, well past the average, the likelihood of frost-damage to fruit is slight. There is nothing much worse than for the blossoms to be forced by a period of warm weather early, for there is only a slim chance that it will continue past the danger limit. It is surprising how late frost may occur,--the last date for killing frost in Pennsylvania is about May 10th on the average, which makes it possible till June.

"The first robins indicate the approach of spring." But certainly not its arrival.

"If the moon rises clear expect fair weather." Right; because if it is summer even the eastern horizon would show the humidity necessary enough to cause a thunderstorm, and in winter the cirrus clouds give several hours' warning. But, again, the wind is the chief factor to be considered.

Proverbs, representing variations of the truth, could be given about every manifestation of the skies as well as about things that were never manifest except in the imagination, for every country has contributed to the volume of weather-lore. But, unfortunately, neither age nor amount of repetition are as good as the truth and they should be discarded if they are false. The way to discard is not to repeat.

The man who desires weather-wisdom should seek it with his eyes. His comparison will be that which he sees with that which he has seen, and he will soon form all the weather axioms he needs for himself. The local Bureau or the Bureau at Washington will answer all his inquiries, cheerfully, promptly, and free of charge. Of course there are things that the Bureau wants to know itself. It is very curious about the higher strata of air. Small balloons have carried very light instruments to an altitude of fifteen miles and brought considerable knowledge to earth, but each bit makes more knowledge imperative.

The cry of "last frontier" hurts the adventurous, the exploring, the woods-loving as no other cry has power to hurt. With the Poles gone and Alaska in harness we are inclined to think that it is all over. We resign ourselves to our trammelling globe,--as the gold-fish do,--forgetting. But there is plenty of interest left. The birds must be brought back. Forests must be made and patrolled, and the air-ocean is still unknown. That, at any rate, has remained unspoiled by man.

The seas have been charted and the mountains have been disemboweled, but the atmosphere is unconquered. More must be known. Squadrons of aeroplanes cannot ride out the gale until their pilots know all about the gale. Until that time there need be no cry of last frontier, for until that time the weather will continue to be our overlord, whose dominions are flaunted before the watcher on the porch and the runner on the trail.

CONDENSATIONS

Look for continued fair weather when:

A gentle wind blows from the west, northwest, or a little south of west.

The sun sets in a cloudless sky.

The sunset is composed of light tints, inclining to red or yellow.

The sunset is followed by a glowing and slow-fading western sky.

The sun sets like a ball of fire (warmer).

The sun rises out of a gray sky.

The clouds are noticeably high for the season.

The clouds rise on the mountains.

The clouds have frequent breaks showing blue sky between.

The puffy cumulus clouds show a lot of white.

The cumulus clouds decrease toward nightfall.

The winter sky is mottled with a northwest wind.

The summer morning fog breaks before ten o'clock.

The dawn is low.

The blue sky has a tendency to show green near the northern horizon (colder).

The sun breaks through a departing thunderstorm and makes a rainbow.

Snow-flurries drift down a north wind (colder).

Cirrus clouds, or others, dissolve, or cirrus have tails down.

Spiders spin on the grass.

There is a moderate dew or frost.

The temperature is normal or colder than normal, other signs being right.

The sky is sown with stars.

The moon rises clear.

The wind blows down mountain ravines after nightfall.

The salt is dry, smoke ascends, birds fly high, and animals act normally.

The barometer rises slowly, or is steady at or above 30.00.

No change need be feared as the anticyclone nears, or for three days after clear conditions are established so long as the wind remains brisk from some westerly quarter. The direction of the wind, the kind of cloud, and the temperature changes are the factors to watch if you have no barometer.

* * * * *

Look for a change toward storms when:

The west wind suddenly drops.

The west wind shifts to south or northeast.

The cirrus clouds appear in well-organized lines.

The cirrus clouds merge into cirro-stratus.

The sky looks like fish scales, so-called mackerel sky.

Light scud drifts across the sky from east to west.

The summer cumulus clouds increase in size as the afternoon proceeds.

Walls grow damp, flies are more of a burden than usual, swallows fly low.

Smoke falls to the ground.

There have been three white frosts.

A halo appears around either the moon or sun.

When sun-dogs appear about the sun, denoting ice-particles in the air.

The summer morning is sultry and the wind variable.

The temperature is much above the normal.

Few stars are visible and those are indistinct. The clouds gather about the mountain tops, or drop down the mountain-sides.

The wind continues to blow up ravines after nightfall.

The sunset is a dull gray, or the sun sets into a livid cloudbank.

The sunrise is a fiery red, and the dawn is high.

The sun gradually is smothered in fine-textured clouds and the wind shifts.

The temperature does not fall at night.

The signs most to be heeded are the shift of wind to a point east of north or south, the gradual filming of the sky with cirrus and cirro-stratus, and the increase of temperature. Of course, the barometer is the best indicator of all.

* * * * *

Look for a change toward clearing when:

The wind shifts from the easterly quarter into the west.

The temperature falls rapidly.

The clouds rise, or break, or lighten perceptibly in color.

Patches of blue sky appear through the rifts in the clouds, wind north.

Raindrops grow smaller after the windshift.

Snowflakes drive less busily, float lazily down, or thin out conspicuously.

Seams appear in the clouds, snow will cease and rain probably.

The thunder and lightning occur only in the eastern quarter.

Permanent clearing will not be effected until the change of the wind to the points on the western half of the compass show that the cyclone has definitely passed to the north or south or over the locality. In winter the cloud covering may move off slowly, but there will be little precipitation after the wind has reached north or west. The bank of cirro-stratus gets thinner and the moon or the sun gradually shines through. In summer clearing is much more abrupt, as is the clouding up. The ability to sense accurately the moment when the weights are shifted and the change to clearing commences takes some observation to acquire, but the advantage is worth it.

* * * * *

Rain (or snow) will fall:

Within five minutes after the arch of the thundercloud is seen to move toward one.

Within five minutes when the curtain of falling drops obscures the landscape to the west of one.

Within a few minutes after the bottoms of cumulus clouds turn from black to gray, letting down visible trailing showers.

Within a short while after the winter sky has become uniform in color.

Within an hour after the pavement-like, but scarcely discernible, thundercloud consolidates along the west, if the wind is from the southwest. If the wind is from the southeast this cloud may take four hours to rise.

From two to eight hours after the sun or moon has vanished behind the cirro-stratus.

From eight to forty-eight hours after the first cirrus is seen, depending upon the distance from the sea and the time of year.

Every little while from southwest showers in the passing of a summer low.

For about eight to twelve hours continuously in a winter storm, and intermittently until the wind swings west.

For a very short while from a thunder cloud rising on a west wind.

For an hour or more from a thundercloud that rises on a southwest or southeast wind.

* * * * *

The temperature will fall when:

A thunderstorm breaks, continuing low if the wind blows from the west after clearing.

Nightfall approaches and the sky is free from clouds.

The mercury remains at the same level during the sunny hours.

A cyclone is departing and the anticyclone moving in.

The wind swings north of east in a storm,--the fall will be gradual.

The wind swings west of south in a storm,--the fall will be sudden.

A snowstorm begins, for a short time only.

A cloudy day clears at sunset.

Snow flurries are seen.

The sky shows green and the clouds look hard.

* * * * *

The temperature will rise when:

A thunderstorm is brewing, or a day or two before a winter cyclone.

After a thunderstorm if another is to follow.

The morning is free from clouds and if it is not the first day of a cold wave.

The wind dips south of west or south of northeast, the former shift bringing the more sudden rise.

The sun sets as a ball of fire, at which one can easily look.

A snowstorm gets under way, unless the wind is swinging toward the north.

A PAGE OF PROBLEMS

One satisfying thing about meteorology is that there is a constantly widening field for conquest. Among the questions that await solution are:

What are the relative densities of clouds?

What is the original atmospheric electricity, its distribution and laws?

What are the causes and nature of precipitation?

Will a 雛 ial ascents on all sides of an atmospheric disturbance discover the mechanism of storms?

What relations are there of solar radiation to our atmosphere?

What influence do lunar tides bear to our weather?

On what does the permanence of the summer lows over the Rockies depend?

These questions are only samples. Many certainties can be attained by merely complete observations over a longer period of time, others by new systems of observations that await a more generous appropriation. Even the upper air investigations on Mt. Weather, Va., have had to be curtailed. The Bureau's record has proved it efficient, of enormous benefit to the country, and deserving of the encouragement instead of the depreciation of every citizen.

WHAT THE WEATHER FLAGS MEAN

In every city the Bureau causes flags to be flown from some prominent place so that a glance may show shippers and everybody who may be concerned at the shortest possible notice just what the approaching weather conditions are.

A plain white flag means fair weather.

A black triangle stands for temperature and is always exhibited with some other flag. Its relative position, either above or below indicates higher or lower temperature. Therefore white flag with the black below means fair and colder. The white flag with the black above means fair and warmer.

A white flag with a black square in the center means a cold wave.

A blue flag means either rain or snow.

The blue with the black above would mean rain or snow and warmer.

The blue with the black below would mean rain or snow and colder.

A blue and white flag means a local shower. The same meanings are attached to the black triangle in connection with the blue and white.

A red triangle indicates a dangerous local storm, is called the information flag meaning that shippers should apply to the Bureau for news of the direction in which the storm is travelling.

A red square with a black center means severe winds.

1. Southwesterly with a white triangle below.

2. Northwesterly with a white triangle above.

3. Northeasterly with a red triangle above.

4. Southeasterly with a red triangle below.

OUR FOUR WORLD'S RECORDS,--AND OTHERS

Maximum Temperature

United States, 134 at Greenland Ranch, Cal., July, 1913.

World, 134 at Greenland Ranch, Cal.

Minimum Temperature

United States, -65 at Miles City, Mont., January, 1888.

World, -98 at Verkhojansk, Siberia.

Absolute Zero of Space

-459 degrees Fahrenheit.

Maximum Annual Precipitation

United States, 167.29 inches at Glenora, Oreg., in 1896.

World, 905.1 inches, Cherrapunji, India, 1861.

Maximum Monthly Precipitation

United States, 71.5 inches at Helen Mine, Cal., January, 1909.

World, 366 inches, Cherrapunji, India, July, 1861.

Maximum 24 Hour Precipitation

United States, 21 inches at Alexandria, La.

Minimum Annual Precipitation

United States, none at Bagdad, Cal., in 1913. (Only 3.93 inches fell at Bagdad during period 1909 to 1913, inclusive.)

Maximum Annual Snowfall

United States, 786 inches at Tamarack, Cal., 1911.

Maximum Monthly Snowfall

United States, 390 inches at Tamarack, Cal., January, 1911.

Maximum Wind Velocity

United States, 186 miles per hour at Mt. Washington, on Jan. 11, 1878. (Much higher velocities have undoubtedly occurred in tornadoes, etc., but have not been susceptible of instrumental measurement.)

THE END

www.ingramcontent.com/pod-product-compliance
Lightning Source LLC
Chambersburg PA
CBHW072255200526
45168CB00016B/1958